MICRO- and NANOSCALE PHENOMENA in TRIBOLOGY

MICRO- and NANOSCALE PHENOMENA in TRIBOLOGY

Edited by
Yip-Wah Chung

CRC Press
Taylor & Francis Group
Boca Raton   London   New York

CRC Press is an imprint of the
Taylor & Francis Group, an **informa** business

CRC Press
Taylor & Francis Group
6000 Broken Sound Parkway NW, Suite 300
Boca Raton, FL 33487-2742

© 2012 by Taylor & Francis Group, LLC
CRC Press is an imprint of Taylor & Francis Group, an Informa business

No claim to original U.S. Government works

First issued in paperback 2017

Version Date: 20110824

ISBN 13: 978-1-138-07235-0 (pbk)
ISBN 13: 978-1-4398-3922-5 (hbk)

Visit the Taylor & Francis Web site at
http://www.taylorandfrancis.com

and the CRC Press Web site at
http://www.crcpress.com

Contents

Preface

Whether dealing with large bearings and gears in a wind turbine or microscale components inside a digital projector, successful operations of these macro- and microdevices depend on understanding and controlling interactions occurring at the micro and the nano scales. This is a field that has benefited a great deal from multidisciplinary collaborations, as evidenced by the numerous international conferences and symposia on the subject over the past twenty years. It is with this spirit that we put together this monograph, which is the result of a short course presented by the National Science Foundation (NSF) Summer Institute on Nanomechanics, Nanomaterials, and Micro/Nanomanufacturing, held in San Diego, California, in April 2010. I hope this monograph represents not a collection of isolated subjects, but a tapestry of the convergence of the multiple science and engineering disciplines and the bridging from the macro to the micro world.

The monograph begins with a short narrative on the evolution of tribology in the micro and nano world. Chapter 2 describes the range of contact conditions spanning the gap between macroscale and nanoscale contacts. Then a primer on macroscale sliding phenomena and how these relate to interfacial film formation and friction performance is presented, followed by a review of instrumentations for examination of microscale sliding contacts. Chapter 3 presents an overview of fundamental continuum treatments of interfacial contact and lubrication under a wide range of conditions, including recent advances in contact simulation. Given the large surface-to-volume ratio in nanoscale materials, structures, and devices, surface forces are destined to be dominant. Chapter 4 gives a thorough account of the nature of surface energies and forces in these structures, as well as adhesion in dry and wet environments. This sets the stage for the next two chapters. Chapter 5 describes how one performs friction measurements at the nano scale and how such friction data can be interpreted, sometimes within the framework of continuum mechanics. Given that magnitudes of surface forces and friction can be modulated by surface topography, Chapter 6 begins with a discussion of various experimental techniques to fabricate micro- and nanotextured surfaces. This is followed by a comprehensive series of results demonstrating how such textures affect adhesion, friction, and wetting.

Tribological properties are affected not only by the surface topography, but also by the environment. Chapter 7 emphasizes the importance of surface chemistry in tribology and reviews some of the environmental effects reported for various tribological interfaces of metals, ceramics, coatings, and solid lubricants. The chapter also presents an in-depth discussion of the effects of alcohol and water vapor on capillary adhesion, friction, and wear of silicon oxide surfaces, followed by examples where environmental effects can be used to mitigate friction and wear. Chapter 8, the final chapter, is on molecular dynamics simulation, from the basic question of what it is and what it can do for tribology, to examples where molecular dynamics simulation has made significant contributions, and to investigations to extend the length and time scales of simulation.

This monograph project is the culmination of efforts of many friends and colleagues. I wish to express my sincere thanks to all the authors for their hard work and cooperation to make this monograph a reality: Allison Shatkin, Jennifer Ahringer, and Andrea Dale of Taylor and Francis for overseeing the publication details, Dr. Ken Chong and Dr. Clark Cooper of the National Science Foundation for their continuing support of the Summer Institute, and Alpana Ranade of Northwestern University for the logistical support of the short course.

Yip-Wah Chung
Northwestern University
March 2011

Editor Biography

Yip-Wah Chung obtained his PhD in physics from the University of California at Berkeley. He joined Northwestern University in 1977. He is currently professor of materials science and engineering and mechanical engineering at Northwestern. His research interests are in surface science, tribology, thin films, and alloy design. He was named Fellow, ASM International; Fellow, AVS; and Fellow, Society of Tribologists and Lubrication Engineers. His other awards include the Ralph A. Teetor Engineering Educator Award from SAE, the Innovative Research Award and Best Paper Award from the ASME Tribology Division, the Technical Achievement Award from the National Storage Industry Consortium (now the Information Storage Industry Consortium), the Bronze Bauhinia Star from the Hong Kong Special Administrative Government, and the Advisory Professor from Fudan University. Dr. Chung served two years as program officer in surface engineering and materials design at the National Science Foundation. His most recent research activities are in infrared reflecting coatings, low-friction surfaces, strong and tough coatings, and high-performance alloys. His favorite hobbies are photography and recreational flying. He holds several FAA ratings, including commercial multiengine instrument, instrument ground instructor, and advanced ground instructor.

Contributors

Rachel Cannara
Center for Nanoscale Science and
  Technology
National Institute of Standards and
  Technology
Gaithersburg, MD

Robert W. Carpick
Mechanical Engineering & Applied
  Mechanics
University of Pennsylvania
Philadelphia, PA

Yip-Wah Chung
Department of Materials Science and
  Engineering
Northwestern University
Evanston, IL

Seong H. Kim
Department of Chemical Engineering
Pennsylvania State University
University Park, PA

Ashlie Martini
School of Engineering
University of California
Merced, CA

Kathryn Wahl
Naval Research Laboratory
Washington, DC

Q. Jane Wang
Department of Mechanical
  Engineering
Northwestern University
Evanston, IL

Min Zou
Department of Mechanical
  Engineering
University of Arkansas
Fayetteville, AR

1 Introduction

Yip-Wah Chung

CONTENTS

1.1 HISTORY

Tribology, the study of friction, wear, and lubrication has become a multidisciplinary endeavor. Historically, mechanical engineering was the home of tribology. Much of the early studies were focused on contact stresses, lubricant film thickness, flash temperatures, and wear modeling, mostly in powertrain and manufacturing components. In lubricated contacts, one can minimize wear by operating under conditions where two sliding surfaces are separated by a lubricant film, with thickness at least three times that of the composite surface roughness to ensure full separation. This is known as the *full-film* or *hydrodynamic lubrication* regime. However, whether driven by economics or the nature of technology never satisfied to be left alone, the performance of mechanical systems continues to be pushed to higher levels—higher loads, higher temperatures, smaller form factors, and lighter structures. As a result, sliding interfaces no longer have the luxury of being separated by a full lubricant film, and direct contact between surfaces often occurs.

The importance of surface composition and testing environment in controlling friction and wear became clear to Don Buckley of the National Aeronautics and Space Administration (NASA) in the 1960s (Buckley 1968a, b). He performed fundamental tribological studies using surface analytical techniques and in situ friction measurements under well-defined environmental conditions. Buckley is truly the pioneer of his time, demonstrating the close interaction between surface science and tribology. Unfortunately, most surface scientists during that period were focusing their attention more on semiconductor surfaces and catalysis than tribology. It wasn't until 1991 that the first tribology symposium was held outside traditional tribology societies. The American Chemical Society organized its first symposium on tribology (Surface Science Investigations in Tribology), involving international speakers from universities, industry, and government laboratories (Chung, Homola, and Street 1992). This symposium brought together researchers from chemistry, chemical engineering, materials science, mechanical engineering, and physics to talk about different aspects of tribology. To the best of my knowledge, this is also the first symposium in which molecular dynamics simulation studies in tribology were presented. This marks the beginning of many such interdisciplinary symposia hosted by the American Chemical Society, American Society of Mechanical Engineers, American Vacuum Society, the Society of Tribologists and Lubrication Engineers, and others.

1.2   IMPACT OF TRIBOLOGY

In the ensuing twenty years, the tribology community has witnessed the convergence of surface science, development of new micro- and nanoscale diagnostic techniques, invention of novel materials, coatings, and lubricants, and demand for higher operating efficiencies and durability in machineries large and small. We have made great strides in improving machine efficiency and durability. An excellent example is the internal combustion engine. Back in the 1970s, one cubic inch of engine displacement on average produced 0.5 hp, and the recommended oil change interval was 3,000 miles. Today, there is no shortage of internal combustion engines giving more than 1.5 hp per cubic inch of engine displacement, and oil change is typically at 5,000-mile intervals or greater. Powertrain and drivetrain components are now much more reliable than before. Passenger cars lasting more than 100,000 miles are now considered more a norm than a rarity.

While tribology cannot claim credit for all the advances of the internal combustion engine and other devices, there have been many numbers cited for the significant economic benefits of proper application and practice of tribology, in the range of a few percent of gross domestic product (GDP). Regardless, benefits of having a reliable piece of tribological hardware go beyond dollars and cents. A crashed hard drive or a stalled car may entail some inconvenience and frustration, but a prematurely worn orthopedic implant can affect someone's health and quality of life. In some instances, a failed piece of tribological hardware can have fatal consequences. Nothing illustrates this better than Alaska Airlines Flight 261, which crashed in the Pacific Ocean on January 31, 2000, about 40 miles west of Malibu, California. All 88 people onboard perished. According to the National Transportation Safety Board report (Aircraft Accident Report NTSB/AAR-02/01), the probable cause of this fatal accident was due to "a loss of airplane pitch control resulting from the in-flight failure of the horizontal stabilizer trim system jackscrew assembly's acme nut threads. The thread failure was caused by *excessive wear* resulting from Alaska Airlines' *insufficient lubrication* of the jackscrew assembly" (emphases added). This is a sober reminder that tribological failure can and will affect people's lives in profound ways.

1.3   THE WAY FORWARD

While performance improvements of conventional machineries can be achieved by evolutionary changes of existing materials, technologies, and practices, the same cannot be said for conventional machineries subjected to more demanding operating conditions, or new mechanical devices that operate under entirely different length scale regimes. In the former situation, one may be dealing with some combination of high flash temperature and high contact stress in a chemically reactive environment, such as may occur in today's advanced high-power-density drivetrain and powertrain components. The most notable examples of the latter are computer disk drives and microelectromechanical systems (MEMS). In these instances, tribological interactions between surfaces depend on phenomena happening at the micro- and nanoscale. In both situations, bulk properties and continuum mechanics alone are no longer adequate to fully describe these interactions. This could be the result of dynamic material transfer from one surface to another, micro- and nanoscale roughness influencing capillary condensation and lubrication, surface chemical reactions, surface forces controlling adhesion, and so on. With the advanced experimental techniques available today, we can study these mechanical and surface chemical phenomena with high degrees of precision, including buried interfaces in some cases. We now have powerful molecular dynamics techniques to simulate micro- and nanoscale tribological phenomena over reasonable length and time scales, providing us with insights and details not accessible before.

Researchers from physical and biological sciences, various disciplines of engineering, and medicine are now working on different aspects of tribology. The success and future of tribology must rely on collaborative efforts across traditional boundaries to gain better understanding of the fundamental micro- and nanoscale interactions, and to apply such knowledge to the design and fabrication of durable tribological components for engineering and biological systems. More than ever, tribology will play a critical role in our quest for a sustainable future, whether by increasing energy efficiency of mechanical systems, enhancing reliability of powertrain systems subjected to frequent start-stops associated with hybrid vehicles, or designing new bearing and gear surfaces for durable wind-turbine operations. The tribology community has much to contribute for a better world!

REFERENCES

Buckley, D. H. and Johnson, R. L. 1968a. The influence of crystal structure and some properties of hexagonal metals on friction and adhesion. *Wear* 11, 405-419.

Buckley, D. H. 1968b. Influence of chemisorbed films of various gases on adhesion and friction of tungsten. *J. Appl. Phys.* 39, 4224–4233

Chung, Y. W., A. M. Homola, and G. B. Street, eds. 1992. *Surface Science Investigations in Tribology: Experimental Approaches*. ACS Symposium Series 485. Washington, DC: American Chemical Society.

NTSB/AAR-02/01, *Loss of Control and Impact with Pacific Ocean Alaska Airlines Flight 261 McDonnell Douglass MD-83, N963AS About 2.7 Miles North of Anacapa Island, California, January 31, 2000*, p. xii, National Transportation Safety Board, Washington DC.

2 Macroscale to Microscale Tribology
Bridging the Gap

Kathryn J. Wahl

CONTENTS

2.1 INTRODUCTION

The past quarter century has brought many changes in our ability to simulate, create, and delicately probe surfaces and structures at ever decreasing scales. Experimentalists have learned to create complex nanostructures, perform ever finer lithography, and pattern single monolayers of graphene. Ultrathin and small structures are readily investigated by tools such as scanning tunneling microscopy (STM) and atomic force microscopy (AFM), which have matured into a broad array of nanoscale surface characterization and imaging techniques widely available on user-friendly commercial platforms. Similarly, these advances provide big opportunities and changes to the field of tribology. Now, nearly every laboratory is equipped with tools that can examine interfaces having contact widths of atomic dimensions with sub-nN force resolution. Where once it was a challenge for molecular dynamics simulations to include a single atom defect in a lattice, modern simulations include

5

thousands of atoms, incorporate amorphous materials, and combine atomic and continuum modeling to tackle larger systems. We now have broad capabilities to examine sliding contacts and adhesion over scales spanning many orders of magnitude, down to the single asperity level.

The focus of this chapter is to treat the topic of experimental tribology, specifically with the aim of examining *what* microscale processes occur in sliding contacts and *how* they impact friction processes. At the macroscopic scale, contacting interfaces are large enough (many microns to mm wide) that we can often literally see what processes are taking place during sliding. At smaller scales, we use the term *microtribology* broadly to both categorize a class of machines or moving assemblies that are small in scale (e.g., micromotors) as well as to describe tribological contacts that fall into the "space between" engineering scale devices and single asperities. Why is this important? Similar to large-scale systems, the design and implementation of small-scale sliding structures depends on understanding and predicting the durability of interfaces, but with the added complication that the interface may be composed of just a few contacting asperities. Do continuum models apply? What about surface effects or wear debris? The surface-to-volume ratio of small devices is also greater as size decreases, and surface forces (like capillary forces) that are not significant contributors in macroscale contacts may suddenly dominate. For this reason, geckos and flies can walk up walls and hang from ceilings using the many microscale hairs on their feet [1,2], while microdevices may realize only limited performance or exhibit complete seizure [3]. There are now many miniature devices in everyday products ranging from accelerometers controlling deployment of airbags in automobiles to inkjet heads in printers and mirror arrays in projectors [4,5]. The failures of these microelectromechanical systems (MEMS) or devices are dominated by adhesion, friction, and wear [5,6]. Thus, it is imperative to develop fundamental understanding of contacting interfaces comprising materials found in microscale systems with the appropriate scales. This is the regime of microtribology.

2.2   WHAT SCIENTIFIC ISSUES AND QUESTIONS SPAN THE GAP?

Perhaps the biggest challenge facing the experimental tribologist is how to devise and perform well-conceived experiments to simply compare friction and wear behavior at different scales. Multiple instruments will likely be used, and it will be necessary to make choices or compromises across the range of loads, contact stresses, sliding speeds, contact dimensions, geometry, and materials. Many of these topics are treated in detail in other chapters comprising this book. Take, for example, the simple experimental configuration of two counterbodies, one spherical and the other planar. This is referred to as sphere-on-flat geometry. This geometry is advantageous in the laboratory for two reasons. First, it minimizes sample alignment problems inherent in cylindrical or flat-on-flat geometries. Further, analytical expressions to evaluate elastic contact parameters (e.g., contact width, deformation, pressure, etc.), widely known as *Hertzian contact mechanics*, are used [7,8]. In the absence of adhesion, the projected contact size, a, is given by

$$a = \left(\frac{3PR}{4E_r} \right)^{1/3}$$

where P is the applied load, R is the radius of the sphere, and E_r is the reduced modulus

$$E_r = \left[\left(1 - v_1^2\right)/E_1 + \left(1 - v_2^2\right)/E_2 \right]^{-1}$$

with E_1, E_2, v_1, v_2 as elastic moduli and Poisson's ratios of the two contacting materials [7]. Ideally, one could change a single parameter at will—for example the contact width—and examine the tribological consequences. Practically, this is very difficult to achieve due to limitations in instrumentation available across scales.

Figure 2.1 shows plots of how changing radius, modulus, and load affect the average contact pressure, or load, divided by the projected contact area, where a is the

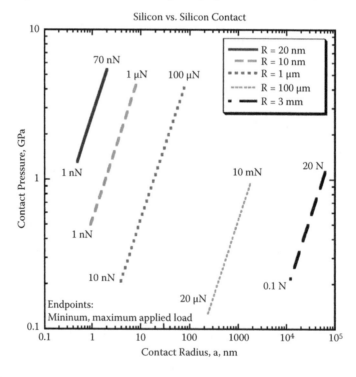

FIGURE 2.1 Hertz contact radius vs. average contact pressure calculated for silicon–silicon contacts for load and radius combinations representative of what can be achieved with a macroscopic pin-on-disk tribometer (broad dashed line at far right), an atomic force microscope (AFM) (solid and medium dashed lines at left of plot), and for a micro-scale tribometer (dotted lines, center). From this plot, it is readily seen that comparisons across lateral scales in contact radius over many orders of magnitude is possible.

radius of a contact of circular cross section ($P/\pi a^2$). This analysis assumes that roughness does not contribute and that the predicted contact area is equal to the true contact
area. First, one can see that in comparing sliding experiments at different scales, it is
a challenge to scale all parameters equally. For example, it would be very difficult to
directly compare measurements made by AFM with a 20-nm radius tip to those made
with a macroscopic tribometer under idealized conditions because the accessible load
ranges and contact pressures are very different. However, with judicious selection of
counterbody radius, a set of experiments might be performed to compare tribological
response over a range of contact dimensions at similar stress ranges.

To illustrate further the challenges in scaling systems, we can look at how
researchers developing MEMS structures [9] have attempted to compare and predict
friction performance with a variety of test geometries and scales. For example, can
friction behavior of a given macroscopic contact predict performance of microscale
devices? What experimental materials and interface parameters are important (load,
sliding speed, time between contacts, capillary forces, roughness, etc.)?

Figure 2.2 shows a summary of macroscopic and microscopic MEMS tribometer
test configurations (upper) and data (lower) for a variety of materials combinations.
The figure summarizes data from an assortment of macroscopic and microscopic
experiments, where static friction coefficient (the force resisting initiation of sliding
motion, relative to the applied load) is compared for various contact geometries,
environments, and material combinations. While general trends for pin-on-disk or
ball-on-disk geometries at higher loads reliably exhibit differences between air and
vacuum environments (lower static friction in vacuum), values span nearly an order
of magnitude. Macroscopic values are generally lower. Researchers could reliably
perform experiments at lower loads and textured surfaces that may provide contacts
more similar to those found in the MEMS devices (e.g., sidewall [11] and nanotractor
[12,13] tribometers).

An additional point to notice regarding the simple pictures presented by Figures 2.1
and 2.2 is that it is not always possible to control or measure the experimental geometry confidently enough to definitively determine contact conditions such as average contact pressure. This is typically due to surface roughness or less-than-ideal
counterface geometries. For example, researchers developing microscale actuators
to make MEMS microtribometers from polycrystalline silicon (polysilicon) may be
able to monitor the applied normal load and resulting static friction coefficient, but
do not necessarily have a confident measure of kinetic friction (the force resisting
sliding motion, relative to the applied load, during sliding), the contact area, or even
how many asperities may be contributing to the contact phenomena. In this case, as
demonstrated in Figure 2.2, researchers have been exploring comparisons between
friction, or sometimes wear [13], measured by microdevices, and those measured
using the same or similar materials using macroscopic tribometers.

Figure 2.3 diagrams the potential to perform experiments with overlap between
macro-, micro-, and nanotribology regimes. The central rectangular region shows
the approximate zone for experiments where the whole instrument and contact is
microscale in dimension (e.g., MEMS device tribometer platforms) [10–14]. The
right side, labeled "bulk," refers to typical pin-on-disk macrotribology experiments, and the left side is the regime typically accessible by AFM instrumentation.

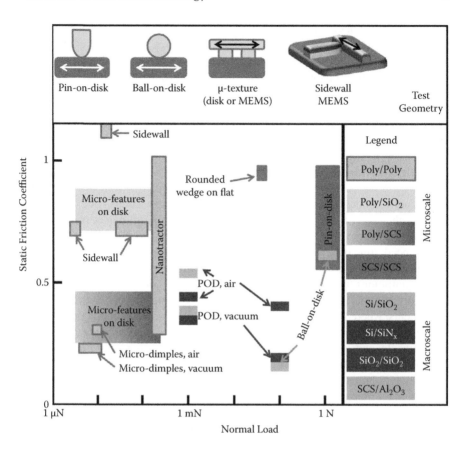

FIGURE 2.2 Comparison of microscale and macroscale static friction coefficient over a broad range of normal loads in air and vacuum environments. Materials in legend text include polysilicon (poly), silica (SiO2), single crystal silicon (SCS), substoichiometric silicon nitride (SiNx), and alumina (Al2O3). Adapted from Alsem, D.H., Dugger, M.T., Stach, E.A., and Ritchie, R.O. 2008. *J. Microelectromech. Sys.* 17, 1144–1154, Table 1, and including data from Lumbantobing, A., and Komvopoulos, K. 2005. *J. Microelectromech. Sys.* 14, 651–663.

In the middle are ovals depicting three regimes that have all been commonly referred to as "microtribology" in the scientific literature. The uppermost region is the regime accessible by commercial *scratch-testing* [15] instruments. In scratch-testing instruments, a sharp diamond tip is moved laterally across the sample (often a coated surface) while the load is increased linearly and the lateral force is monitored. The middle regime shows the approximate range of sliding nanoindentation experiments (with applied loads controlled between ~uN to ~mN) where the counterbody tips are much blunter than those used for scratch testing. On average, the blunter tip substantially reduces the contact stresses to GPa or lower. Contact stresses can be reduced further by employing larger counterface radii (e.g., 100 µm to mm scale) on a tribometer capable of controlling applying very low loads during sliding.

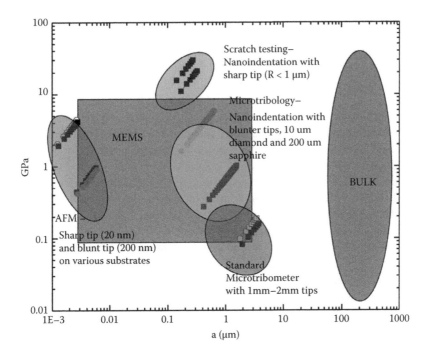

FIGURE 2.3 Plot similar to Figure 2.1 with examples of the calculated Hertzian contact size and contact pressure for typical AFM experiments (leftmost oval), macroscopic tribometry (*bulk*, rightmost oval), and MEMS-scale experiments (rectangle). The middle three "microtribology" regimes—center ovals depicting nanoindentation scratch testing, nanoindenter-based experiments with blunter tips, and mm-scale sphere microtribometers—fall in the middle range. From the plot, one can see that multiple instrumentation configurations can probe similar contact widths across many orders of magnitude in average contact stress, or across many orders in contact dimension with similar contact stresses.

To summarize, there are three regimes commonly referred to in the scientific literature as *micro*tribology regimes. These include (1) experiments where everything is microscale—employing actual MEMS structures, (2) experiments where the loads and stresses are small and the sliding distances big (microtribometers, which are often commercially misrepresented as *nanotribometers*), and (3) experiments where the contact and sliding dimensions are small, but the stresses might be big (nanoindentation-based sliding experiments). All of these provide the experimenter with a rich opportunity to explore the gap between nano- and macroscale contacts.

Ultimately, one wants to know what phenomena are responsible for the observed mechanical behavior (friction, wear, adhesion, etc.) in microscale devices and contacts. Why is the friction low or high? What initiates wear? Have the materials in the contacting interface changed chemically or mechanically? Can they be lubricated, and what works best? How does scaling affect lubrication by solids or liquids? Is anything deforming or moving, and if so, how—by rolling, shearing, plowing, or fracture? How can we best examine and develop a working knowledge of microscale contacts?

Some of these questions can be answered by straightforward experimental measurements and perhaps clever selection of test conditions. Understanding sliding phenomena at a minimum requires analytical evaluation of the sliding contacts after testing—one might call this the *CSI* (crime scene investigator) or forensics approach—examine the sliding interfaces after testing and attempt to figure out what happened. There are many analytical tools with high spatial resolution that can evaluate surface morphology, wear debris particle sizes and chemical composition, wear track depths, surface chemistry changes, materials deformation, lubricant degradation, and so on. These include scanning electron microscopy (SEM) and energy dispersive x-ray (EDX) analysis, scanning Auger microprobe (SAM), x-ray photoelectron spectroscopy (XPS), micro-Raman spectroscopy, Fourier Transform Infrared (FTIR) spectroscopy, atomic force microscopy (AFM), optical profilometry, and nanoindentation.

Unfortunately, many of the questions listed previously cannot easily be answered after the fact, when sliding has stopped and the contacts have been separated. For macroscopic contacts, one solution has been to devise methods to monitor the events in the sliding contact *during* sliding; these approaches are termed *in situ*, from Latin meaning "in place," and are distinct from studies done after sliding contacts have been stopped or separated, ex situ. The in situ approaches (typically involving optical, electronic, and/or spectroscopic monitoring of sliding contacts) have proved very valuable in demonstrating how macroscopic contacts are lubricated, and also how they fail [16,17].

The remainder of this chapter will address the issue of bridging the gap between macroscale and microscale tribology. First, we will discuss some examples of how in situ tribology can provide insight into contact phenomena at macroscopic scales. While sliding processes may not all translate between scales, surely some will, and a working knowledge of what can happen in sliding contacts at larger scales may prove useful as contact dimensions are reduced. As scale decreases (e.g., nanoscale to atomic scale), continuum phenomena may break down. This is a fascinating topic (see e.g., [18,19]) but beyond the scope of this chapter. Following the macroscopic examples, we will broadly address some of the most typical instrumentations available to explore sliding contacts falling in the microtribology regime.

2.3   IN SITU TRIBOLOGY: WHAT'S HAPPENING IN THE BURIED SLIDING INTERFACE?

Interfacial processes play a key role in the friction and wear performance of materials and devices. Studying sliding interfaces, which are both hidden from view and moving relative to each other, is challenging. Most of our understanding of friction and wear has come from ex situ studies of separated contacts. However, many of the processes controlling tribological response of an interface involve dynamic processes (material deformation, plowing, chemical transformation, rheology, etc.). Interfacial processes often involve changes to the original materials, so that understanding the structure and processing of the parent materials is not enough to predict performance of a material pair. To understand how interfacial changes play a role in friction and wear performance of a sliding couple, tribologists have applied a wide variety of surface analytical tools to examine buried sliding interfaces at scales from macroscopic contacts to single asperities. These techniques include optical microscopy, interferometry, high speed videography, chemical spectroscopy (Raman, FTIR), and electrical contact resistance studies, among many in a rich history of sliding contact physics (see e.g., Bowden and Tabor [20]). Furthermore, engineers are developing bearings with windows and sensors for real time monitoring of performance.

There are many examples of research occurring across a broad spectrum of materials science, physics, and chemistry that relate to examining the science of buried interfaces, and the phenomena within, which control bearing performance. We will discuss two here, with a twofold purpose: (1) to demonstrate the power and utility of in situ approaches to understanding friction processes, and perhaps more importantly, (2) to inspire researchers working at ever smaller scales to design experiments and devise new methods to probe the dynamic interfacial events including friction, wear, and chemistry that control sliding contacts.

At the Naval Research Laboratory, we have used a very simple geometry that combines visual observation through a transparent, stationary counterface with spectroscopic analysis using a Raman microprobe [21–25]. In this way, the interfacial films and debris formed during sliding can be monitored, quantified, and recorded in real time, and compared to friction performance. Our objective has been to demonstrate what processes influence and control friction in contacts lubricated with solid lubricant coatings such as molybdenum disulfide, diamond-like carbon, boric acid, as well as nanocomposites of these and other materials.

2.3.1 WHAT SEPARATES THE SURFACES?

Thin, soft films are often employed as solid lubricants in regimes where liquid lubricants would not perform well [20,26], such as space/vacuum, or high-temperature environments. Also, for small contacts, capillary forces may be significant and either gas phase [see Chapter 7] or solid phase lubricants may be required. Examples of typical solid lubricants are graphite and the layered dichalcogenides MoS_2, WS_2, and WSe_2, thin soft metal films (Ag, Au, etc.), boric acid, diamond-like carbon (DLC), and nanocomposites combining these materials. At macroscopic sliding scales, the lubrication mechanism for all of these materials is similar in one way—an interfacial film is formed, transfers to the stationary counterbody, and separates and protects the sliding surfaces. The interface can be transformed chemically or mechanically (often called *tribochemistry* or *tribomechanics*): the interfacial film may be thin or thick, chemical or phase transformations can occur, crystalline phases may be oriented or reoriented with respect to the sliding plane, and rheology or movement within the interfacial film may take place.

To answer the question posed previously, we have monitored the interfacial film formation and thickness using several optical methods involving either absorbance [22,23] or interferometry [27]. For the latter approach, interference fringes (e.g., Newton's rings) outside the periphery of the contact were used to monitor and quantify the interfacial film thickness during sliding (see Figure 2.4). The fringe

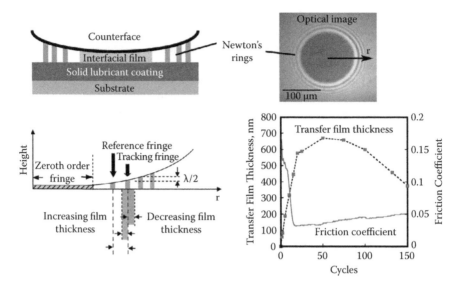

FIGURE 2.4 The upper left diagram shows the cross section of an interfacial film formed from a solid lubricant coating, and the location of interference fringes outside the contact periphery. The upper right photomicrograph from an actual contact shows the interference fringes as viewed from above the contact, before sliding has started (0 sliding cycles). The lower left diagram shows how the movement of the interference fringes can be used to quantify the changes in interfacial film thickness by monitoring the relative height of the spherical counterbody to the coating surface below. The data at the lower right show an example of how the interfacial transfer film thickness changes with sliding cycle, and the corresponding friction coefficient. Adapted from Wahl, K.J., Chromik, R.R., and Lee, G.Y. 2008. *Wear* 264, 731–736.

separation is ~270 nm with white light illumination. As interfacial film thickness changes, the position of the interference fringes changes, moving inward as films thicken, or outward as they become thinner (see diagram at lower left in Figure 2.4). A photomicrograph showing the contact interface before sliding (viewed from the top, through the transparent hemispherical counterface that remains stationary during sliding) is shown at the upper right. The resolution of the technique is limited by camera resolution and fringe sharpness; practically, the technique is limited by debris building up around the contact periphery and obscuring the rings. The lower right portion of Figure 2.4 shows an example for a solid lubricated contact showing that the initial drop in friction is directly correlated with buildup of an interfacial film during the first 20 sliding cycles [27].

2.3.2 Where Is Sliding Taking Place?

From visual analysis of macroscopic sliding contacts like those shown in Figure 2.4, we have observed that a majority of the sliding is accommodated by *interfacial sliding*—that is, between the interfacial film and the coating (wear track) below. Under some conditions, both *interfacial* and *intrafilm sliding* (motion within the interfacial film) take place. Figure 2.5 illustrates this with examples of (a) optical microscopy of interfacial films, (b) diagrams of interfacial sliding without (left) and with (right) intrafilm sliding, and (c) friction coefficient observed during sliding. In this example, Dvorak et al. [24] demonstrated that for a solid lubricating film composed of Pb-doped MoS_2, the sliding was essentially 100% interfacial in dry environments (low relative humidity in air). In contrast, when the humidity was increased, we observed a combination of interfacial and intrafilm sliding, along with

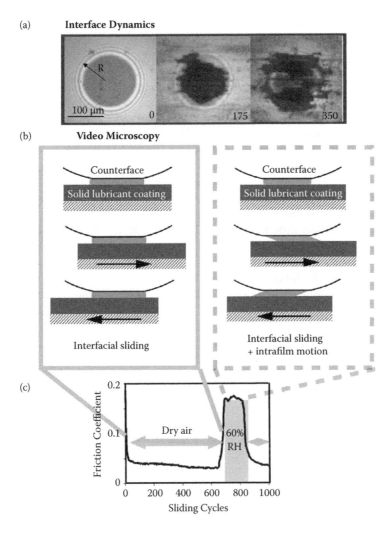

FIGURE 2.5 (a) Sample photomicrographs of interfacial films formed during sliding. (b) Side-on view diagrams of a sliding contact undergoing interfacial sliding (left) and interfacial sliding with intrafilm motion (right). (c) Plot of friction coefficient vs. sliding cycles showing sharp friction rise with addition of ambient humidity to the test environment. See text for further explanation. Adapted in part from Dvorak, S.D., Wahl, K.J., and Singer, I.L. 2007. *Tribol. Lett.* 28, 263–274.

higher friction, Figure 2.5(c). This observation begged the question, "Which came first, the chicken or the egg?"—interfacial film motion or high friction—and why?

Through a series of experiments [24], Dvorak et al. were able to demonstrate that humidity increased the interfacial shear stress to the interface. However, the transfer film did not begin to exhibit intrafilm shear until after it weakened upon exposure to the increased humidity (e.g., the interfacial film strength became lower than the interfacial shear stresses acting on it). Thus, friction rose due to the humidity increase, and the shearing of the interfacial film resulted from its being weaker than the forces acting at the interface during sliding.

2.3.3 What Is Really Happening in the Contact?

It is important to note that so far we have only observed 2 of the 20 or so predicted *velocity accommodation modes*—interfacial sliding and intrafilm shear—predicted by Godet and coworkers [28–30]. We infer that those two modes are the dominant sliding accommodation modes for solid lubricants. However, the limitations of what one could expect to see from a macroscopic experimental setup with optical microscopy are significant. We can confirm that large particles are not rolling (they'd be moving at half the sliding speed), but cannot see microstructural reorientation or clearly define exactly how many atomic layers are involved in the sliding itself; that is, exactly how thin is the sliding interface?

Recent microscale experiments have begun to address these important questions. Ideally, one would want to directly image atomic-scale motion within sliding interfaces, or at least motion of crystalline structures. In cross section, this can be done using high-resolution transmission electron microscopy (HR-TEM). Several groups are bringing in situ TEM to the problem of interfacial tribology [31]; this approach is in its infancy and has enormous potential to address the key questions of how and where interfaces slide, as well as what chemical and mechanical processes occur during sliding. For example, how exactly do lamellar solid lubricants accommodate sliding motion? Does it require one lamella or two? More? While our macroscopic experiments indicate that the "deck of cards" model is naive and does not reflect sliding accommodation [16,17,21–25], perhaps this approach will reveal just how many layers are acting to allow slip, as well as how it works. As an intermediate step toward this goal, Hu and co-workers [32] have shown that by using focused ion beam (FIB) milling to cross-section a microscale contact (with the sliding counterbody in place), sliding contact against a crystalline coating of WSe_2 resulted in transformed surface microstructure, with both reorientation and recrystallization of the WSe_2 layers. Interestingly, some noncontacting regions of the frozen interface contained no observable interfacial film.

To scale down lubrication schemes, researchers and engineers will need to determine how lubricants work at different scales. Is contact pressure a key variable? Do interfacial film thicknesses scale with contact width? Can small contacts even sustain a transferred film? Many significant and interesting questions remain to be addressed as scientists develop and improve experimental apparatuses capable of providing high-quality data at submicron size scales.

2.4 MICROTRIBOLOGY INSTRUMENTATION: APPROACHES

There is a wide variety of instrumentation available to examine sliding contacts with microscopic dimensions. Broadly, the instruments include AFM, nanoindenters, dedicated microtribometers, and MEMS tribometers. As described earlier in this chapter and shown in Figure 2.3, these platforms span a range of accessible contact sizes, contact stresses, sliding speeds, and stroke lengths.

2.4.1 AFM

AFM friction measurements are covered elsewhere in this book (see Chapter 5), and will not be treated in detail here. Briefly, larger AFM contacts have been obtained through the method introduced by Ducker et al. [33], who used microsphere-tipped cantilevers to create larger tip radii for adhesion experiments (see also [34]). Microspheres (also called *colloidal particles*) used for such studies range from AFM with silica, polymer, or other microsphere-tipped cantilevers [35] and can be chemically functionalized with organic monolayers for additional interfacial control [36]. Early work by Liu et al. [37] demonstrated the potential for colloidal probe AFM to explore velocity-dependent friction of thin molecular films. With appropriate care and calibration, sliding speed can be varied over several orders of magnitude (10s of nm/s to 10s of μm/s), and friction forces from nN to μN accessed with proper cantilever selection and calibration. Stroke lengths (sliding distances) are limited by the scanner size, typically to at most ~100 μm, but generally far smaller scan widths are used. Use of microscale spheres increases both the chance that the tip–sample interaction will be dominated by effects of adhesion and roughness (see, e.g., [38]). Again, the reader is referred to Chapter 5 in this volume.

2.4.2 LOW-LOAD TRIBOMETERS

Several manufacturers make what can be classed as ultralow-load tribometers, with minimum applied loads in the 1–10s of μN range and ability to run unidirectional (rotating pin-on-disk) or reciprocating sliding tests with linear stroke lengths of 10s to 100s of microns (see, e.g., [39–42]), and are well-suited for measurements on compliant systems [40–42]. Often, the instruments are described as *nanotribometers*, which to a degree confuses the naming classification for friction testing apparatuses, particularly since the counterface radii are typically quite large (mm scale). Other low-load tribometers have also been reported in the literature, including one based on a design similar to an AFM (e.g., Dvorak et al. [43]). Generally, these instruments are designed to work in a manner similar to macroscopic tribometers at fixed load for a proscribed number of cycles or until friction rises above a set point. Some instruments have the capability of monitoring the change in deflection of the load transducer with sliding.

2.4.3 NANOINDENTATION-BASED TRIBOMETERS

Some nanoindentation instruments that were originally designed and intended for scratch testing can be used to perform reciprocating sliding tests at low applied loads.

Schiffmann et al. [44,45] described reciprocating sliding studies of DLC coatings using hemispherically tipped conical diamond indenters. Before the experiments, the tip was scanned over the surface to record the unworn surface profile; this could be as wide as 10–20 microns. Then, a set of up to 50 cycles was performed with a set loading profile that could either be constant or ramped to a new load between reciprocating cycles using a sliding length about 10% shorter than the original surface scan. After sliding, the tip was brought over the original surface scan region. This provided two key pieces of information: (1) a reference for drift of the sample vertical position during the experiment (from the regions outside the sliding region), and (2) an estimate of the wear track depth, which could be different from the initial surface due to either wear or plastic deformation.

So et al. [46] used this approach to examine the origins of friction hysteresis of tilted nanofibrillar parylene films, as shown in the top two plots of Figure 2.6. Very little drift was observed (a few 10s of nm at low contact depths) and could be corrected after scanning. Furthermore, during friction scanning at different applied loads (hysteresis loops in the upper plots), clear hysteresis was observed for sliding experiments where the tip was moved parallel to the tilt direction of the parylene film (center left micrograph). In contrast, very little depth hysteresis was observed for films when sliding was perpendicular to the tilt orientation (center right micrograph). Furthermore, during experiments the depth of the contact is measured continually. From the data, friction as well as either wear or contact mechanics information (load/depth) can be gleaned from the microscale sliding experiments. The experiments of So et al. demonstrated friction anisotropy (Figure 2.6, lower plots) that was attributed to contact area changes with sliding orientation (the tip penetrated deeper into the sample when sliding with the fiber nap, increasing the contact area and friction force magnitude).

For additional examples, the reader is referred to Schiffmann et al. [44,45], who explored the contact mechanics of plowing in microscale contacts to DLC coatings, Gao et al. [47], who investigated microslip, and Chromik and coworkers [48,49], who employed similar approaches to explore microtribology and wear of Au-MoS_2 solid lubricant coatings.

2.4.4 MEMS TRIBOMETER PLATFORMS

A number of groups have developed MEMS platforms capable of reporting (directly or indirectly) friction performance of the device itself. This is significant in that the friction or wear behavior of a working device can be monitored in situ, in the desired test or lubrication environment. Given the difficulty of interpreting and projecting MEMS device friction performance based on the broad effort to map static friction response shown in Figure 2.2, an on-chip friction capability is an important accomplishment. Examples of MEMS tribometers include the Berkeley micromachine [50] and the Sandia sidewall tribometer [11]. A similar MEMS device has been fabricated in China [51]. Van Spengen [14] developed a MEMS tribometer capable of reporting friction. Again, the MEMS tribometer platforms significantly enhance the ability of researchers to test lubrication strategies and chemical transformation [52–54] (see also Chapter 7 in this volume) as well as contact conditions under actual device operation.

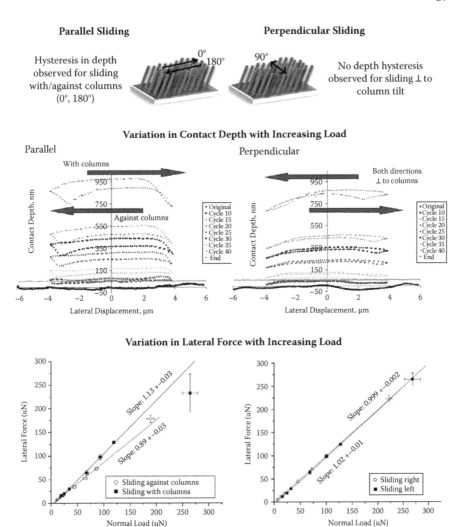

FIGURE 2.6 Examples showing the capability of nanoindenter-based sliding experiments to address not only friction but also contact mechanics questions. The figures at the top show the tip-sample depth during sliding for experiments with a 20-micron radius indenter sliding against tilted, nanofibrillar parylene films in a reciprocating configuration. The lateral displacement vs. depth curves showed clear hysteresis (see text) when sliding with/against the nanofiber tilt direction. Lateral forces and friction coefficient (estimated from the slope of the linear curves in the bottom two plots) also exhibited hysteresis with fiber tilt direction. This is a good example of how using instruments with capabilities beyond lateral force sensing (as in most macroscopic tribometers) can add significant experimental information to test continuum contact mechanics predictions. Figure portions adapted from So, E., Demirel, M.C., and Wahl, K.J. 2010. *J. Phys. D* 43, 045403.

2.4.5 OTHER APPROACHES

Finally, one can benefit by combining several instruments or approaches simultaneously to measure two independent parameters from the same contact. This has been employed extensively in the AFM and nanomechanics communities through oscillatory techniques; by providing a small, linear perturbation to the contact either laterally or normal to the surface using force or displacement, one can measure the contact stiffness from the differential force/displacement ($\Delta F/\Delta x$) response [55]. From this information, the contact size can be inferred and compared to contact mechanics predictions, or used to derive materials properties of the contacting materials. It is also possible to combine two oscillatory techniques (contact stiffness with a nanoindenter tip against a quartz crystal microbalance monitoring frequency shift), operating in different bandwidths and orientations, to evaluate tip/sample contact mechanics [56,57]. Borovsky and coworkers [58] have also explored the nature of slip onset with a similar experimental configuration.

2.5 FUTURE OUTLOOK

Numerous opportunities exist to broaden our understanding of microscale sliding interfaces and the processes that control their performance. This chapter has outlined a number of fundamental questions and approaches to examine the nature of small-scale sliding contacts. By combining an understanding of macroscopic phenomena with new approaches at the microscale, one can hypothesize and test the behavior of ever smaller contacts and interfaces. Most exciting of all, the promise of observing in real time how micro- or even nanoscale contacts slide past each other and accommodate shear is now a reality with the development of tools combining atomic scale visualization with scanned probe microscopy.

ACKNOWLEDGMENTS

This work was funded by the Office of Naval Research and the Basic Research Program of the Naval Research Laboratory. I would like to acknowledge my current and former colleagues Irwin Singer, Bob Bolster, Larry Seitzman, Derren Dunn, David Dvorak, Tom Scharf, Jenny Heimberg, Gun Lee, Colin Baker, Rich Chromik, Nimel Theodore, Bill Unertl, Rich Colton, Syed Asif, Donna Ebenstein, and Eric So for their contributions to interface science that informed this work.

REFERENCES

1. Autumn, K., Liang, Y.A.. Hsieh, S.T., Zesch, W., Chan, W.P., Kenny, T.W., Fearing, R., and Full, R.J. 2000. *Nature* 405, 681–685.
2. Huber, G., Mantz, H., Spolenak, R., Mecke, K., Jacobs, K., Gorb, S.N., and Arzt, E. 2005. *Proc. Nat. Acad. Sci.* 102, 16293–16296.
3. Maboudian, R., and Howe, R.T.J. 1997. *Vac. Sci. Technol. B* 15, 1–20.
4. Spearing, S.M. 2000. *Acta. Mater.* 48, 179–196.

5. Alsem, D.H., Dugger, M.T., Stach, E.A., and Ritchie, R.O. 2008. *J. Microelectromech. Sys.* 17, 1144–1154.
6. Romig, A.D., Dugger, M.T., and McWhorter, P.J. 2003. *Acta Mater. 51*, 5837–5866.
7. Johnson, K.L. 1985. *Contact Mechanics.* Cambridge, UK: Cambridge University Press.
8. Johnson, K.L. 1982. *Proc. Inst. Mech. Eng.* 196, 363–378.
9. Maboudian, R., Ashurst, W.R., and Carraro, C. 2002. *Tribol. Lett.* 12, 95–100.
10. Lumbantobing, A., and Komvopoulos, K. 2005. *J. Microelectromech. Sys.* 14, 651–663.
11. Senft, D.C., and Dugger, M.T. In *Proc. SPIE- Int. Soc. Opt. Eng. 3224,* 31–38.
12. de Boer, M.P., Luck, D.L., Ashurst, W.R., Maboudian, R., Corwin, A.D., Walraven, J.A., and Redmond, J.M. 2004. *J. Microelectromech. Syst.* 13, 63–74.
13. Flater, E.E., Corwin, A.D., de Boer, M.P., and Carpick, R.W. 2006. *Wear* 260, 580–593.
14. van Spengen, W.M., and Frenken, J.W.M. 2007. *Tribol. Lett.* 28, 149–156.
15. Bull, S.J., and Berasetugui, E.G. 2006. *Tribol. Int.* 39, 99–114.
16. Wahl, K.J., and Sawyer, W.G. 2008. *MRS Bull.* 33, 1159–1167.
17. Singer, I.L., Dvorak, S.D., Wahl, K.J., and Scharf, T.W. 2003. *J. Vac. Soc. Technol. A* 21, S232–S240.
18. Luan, B., and Robbins, M.O. 2006. *Phys. Rev. E* 74, 026111.
19. Hurtado, J.A., and Kim, K.S. 1989. *Proc. Roy. Soc. London A* 455, 3363–3384.
20. Bowden, F.P., and Tabor, D. *The Friction and Lubrication of Solids.* Oxford, UK: Clarendon.
21. Dvorak, S.D., Wahl, K.J., and Singer, I.L. 2002. *Tribol. Trans.* 45, 354–362.
22. Scharf, T.W., and Singer, I.L., 2003. *Tribol. Lett.* 14, 137–145.
23. Scharf, T.W., and Singer, I.L., 2003. *Thin Solid Films* 440, 138–144.
24. Dvorak, S.D., Wahl, K.J., and Singer, I.L. 2007. *Tribol. Lett.* 28, 263–274.
25. Chromik, R.R., Winfrey, A.L., Luning, J., Nemanich, R.J., and Wahl, K.J. 2008. *Wear,* 265, 477–489.
26. Singer, I.L. 1992. In *Fundamentals of Friction: Macroscopic and Microscopic Processes,* ed. Singer, I.L., and Pollock, H.M., 237–261. Dordrecht: Kluwer.
27. Wahl, K.J., Chromik, R.R., and Lee, G.Y. 2008. *Wear* 264, 731–736.
28. Godet, M. 1984. *Wear* 100, 437–452.
29. Berthier, M., Vincent, L., and Godet, M. 1988. *Wear* 125, 25–38.
30. Godet, M. 1990. *Wear* 136, 29–45.
31. Marks, L.D., Warren, O.L., Minor, A.M., and Merkle, A.P. 2008. *MRS Bulletin* 33, 1168–1173.
32. Hu, J.J., Wheeler, R., Zabinski, J.S., Shade, P.A., Shiveley, A., and Voevodin, A.A. 2008. *Tribol. Lett.* 32, 49–57.
33. Ducker, W.A., Senden, T.J., and Pashley, R.M. 1991. *Nature* 353, 239–241.
34. Butt, H.J. 1991. *Biophys. J.* 60, 1438–1444.
35. Butt, H.J., Capella, B., and Kappl, M. 2005. *Surf. Sci. Reports* 59, 1–152.
36. Noy, A., Vezenov, D.V., and Lieber, C.M. 1997. *Ann. Rev. Mater. Sci.* 27, 381–421.
37. Liu, Y.H., Evans, D.F., Song, Q., and Grainger, D.W. 1996. *Langmuir* 12, 1235–1244.
38. Jones, R., Pollock, H.M., Cleaver, J.A.S., and Hodges, C.S. 2002. *Langmuir* 18, 8045–8055.
39. Gitis, N., Vinogradov, M., Hermann, I., and Kuiry, S. 2007. In *Fundamentals of Nanoindentation and Nanotribology IV,* Proceedings of the 2007 MRS Fall Meeting, Boston, MA, Vol. 1049, 3–8. Warrendale, PA: Materials Research Society.
40. Rennie, A.C., Dickrell, P.L., and Sawyer, W.G. 2005. *Tribol. Lett.* 18, 499–504.
41. Dickrell, P.L., Sinnott, S.B., Hahn, D.W., Raravikar, N.R., Shadler, L.S., Ajayan, P.M., and Sawyer, W.G. 2005. *Tribol. Lett.* 18, 59–62.
42. Dunn, A.C., Zaveri, T.D., Keselowsky, B.G., and Sawyer, W.G. 2007. *Tribol. Lett.* 27, 233–238.
43. Dvorak, S.D., Woodland, D.D., and Unertl, W.N. 1998. *Tribol. Lett.* 4, 199–204.

44. Schiffmann, K.I., and Hieke, A. 2003. *Wear* 254, 565–572.
45. Schiffmann, K.I. *Wear* 265, 1826–1836.
46. So, E., Demirel, M.C., and Wahl, K.J. 2010. *J. Phys. D* 43, 045403.
47. Gao, Y.F., Lucas, B.N., Hay, J.C., Oliver, W.C., and Pharr, G.M. 2006. *Scripta Mat.* 55, 653–656.
48. Stoyanov, P., Fishman, J.Z., Lince, J.R., and Chromik, R.R. 2008. *Surf. Coat. Technol.* 203, 761–765.
49. Stoyanov, P., Chromik, R.R., Goldbaum, D., Lince, J.R., and Zhang, X. 2010. *Tribol. Lett.* 40, 199–211.
50. Lim, M.G., Chang, J.C., Schultz, D.P., Howe, R.T., and White, R.M. 1990. In *Micro Electro Mechanical Systems, Proc. IEEE 3rd MEMS Workshop, An Investigation of Micro Structures, Sensors, Actuators, Machines, and Robots*, 82–88. IEEE, New York.
51. Wu, J., and Wang, S. J. 2008. *Microelectromech. Sys.* 17, 921–933.
52. Kim, S.H., Asay, D.B., and Dugger, M.T. 2007. *nanoToday* 2, 22–29.
53. Asay, D.B., Dugger, M.T., and Kim, S.H. 2008. *Tribol. Lett.* 29, 67–74.
54. Barnette, A.L., Asay, D.B., Olhausen, J.A., Dugger, M.T., and Kim, S.H. 2010. *Langmuir* 26, 16299–16304.
55. Pethica, J.B., and Oliver, W.C. 1987. *Phys. Scripta* T19A, 61–66.
56. Borovsky, B., Krim, J., Syed Asif, S.A., and Wahl, K.J. 2001. *J. Appl. Phys.* 90, 6391–6396.
57. Ellis, J.S., and Hayward, G.L. 2003. *J. Appl. Phys.* 94, 7856–7867.
58. Borovsky, B., Booth, A., and Manlove, E. 2007. *Appl. Phys. Lett.* 91, 114101.

3 A Continuum Overview of Contact and Lubrication

Q. Jane Wang

CONTENTS

A surface pair under contact and relative motion with or without lubrication forms a tribological interface. Such an interface is found in nearly all types of machines. Contact and relative motion of surfaces make friction and wear inevitable. Theories of contact and lubrication are essential to understanding phenomena related to friction and wear and to developing energy-efficient, robust tribological interfaces. This chapter presents an overview of the fundamentals of contact and lubrication theories together with examples of recent progress in lubrication theory inspired by advances in contact simulation. Challenges and opportunities are also briefly discussed.

3.1 FUNDAMENTALS OF CONTACT MECHANICS

Surface contact analysis methods are based on the classic potential theory (Love 1929, 1952; Johnson 1985). The surface of a half space may be under normal and tangential loading. When area S is subjected to load, the normal pressure, $p(\xi, \eta)$, and tangential tractions in two orthogonal directions, $q_x(\xi, \eta)$ and $q_y(\xi, \eta)$, at a point,

(ξ, η), are the general form of surface tractions. A few potential functions that satisfy the Laplace equation are defined and used for contact elasticity (Johnson, 1985; Hills et al. 1993) with the shear modulus, $G_s = E/[2(1+v)]$, and a couple of geometrical parameters, $\Omega = z\ln(R + z) - R$, $R^2 = (x-\xi)^2 + (y - \eta)^2 + z^2$.

$$F_1 = \iint_S q_x(\xi, \eta)\Omega d\xi d\eta \tag{3.1}$$

$$G_1 = \iint_S q_y(\xi, \eta)\Omega d\xi d\eta \tag{3.2}$$

$$H_1 = \iint_S p(\xi, \eta)\Omega d\xi d\eta \tag{3.3}$$

$$F = \frac{\partial F_1}{\partial z} = \iint_S q_x(\xi, \eta)\ln(R + z)d\xi d\eta \tag{3.4}$$

$$G = \frac{\partial G_1}{\partial z} = \iint_S q_y(\xi, \eta)\ln(R + z)d\xi d\eta \tag{3.5}$$

$$H = \frac{\partial H_1}{\partial z} = \iint_S p(\xi, \eta)\ln(R + z)d\xi d\eta \tag{3.6}$$

Using the potentials above, one can define:

$$\psi_1 = \frac{\partial F_1}{\partial x} + \frac{\partial G_1}{\partial y} + \frac{\partial H_1}{\partial z} \tag{3.7}$$

$$\psi = \frac{\partial \psi_1}{\partial z} = \frac{\partial F}{\partial x} + \frac{\partial G}{\partial y} + \frac{\partial H}{\partial z} \tag{3.8}$$

The deformations at a point (x, y, z) in the half-space solid are found by Love (1952) as follows:

$$u_x = \frac{1}{4\pi G_s} \left\{ 2 \frac{\partial F}{\partial z} + \frac{\partial H}{\partial x} + 2v \frac{\partial \psi_1}{\partial x} - z \frac{\partial \psi}{\partial x} \right\}$$

(3.9)

$$u_y = \frac{1}{4\pi G_s} \left\{ 2 \frac{\partial G}{\partial z} - \frac{\partial H}{\partial y} + 2v \frac{\partial \psi_1}{\partial y} - z \frac{\partial \psi}{\partial y} \right\}$$

(3.10)

$$u_z = \frac{1}{4\pi G_s} \left\{ \frac{\partial H}{\partial z} + (1 - 2v)\psi - z \frac{\partial \psi}{\partial z} \right\}$$

(3.11)

Stresses at point (x, y, z) in the half-space solid can be easily expressed as:

$$\sigma_{xx} = \frac{2vG_s}{1 - 2v} \left\{ \frac{\partial u_x}{\partial x} + \frac{\partial u_y}{\partial y} + \frac{\partial u_z}{\partial z} \right\} + 2G_s \frac{\partial u_x}{\partial x}$$

(3.12)

$$\sigma_{yy} = \frac{2vG_s}{1 - 2v} \left\{ \frac{\partial u_x}{\partial x} + \frac{\partial u_y}{\partial y} + \frac{\partial u_z}{\partial z} \right\} + 2G_s \frac{\partial u_y}{\partial y}$$

(3.13)

$$\sigma_{zz} = \frac{2vG_s}{1 - 2v} \left\{ \frac{\partial u_x}{\partial x} + \frac{\partial u_y}{\partial y} + \frac{\partial u_z}{\partial z} \right\} + 2G_s \frac{\partial u_z}{\partial z}$$

(3.14)

$$\tau_{xy} = G_s \left\{ \frac{\partial u_x}{\partial y} + \frac{\partial u_y}{\partial x} \right\}$$

(3.15)

$$\tau_{yz} = G_s \left\{ \frac{\partial u_y}{\partial z} + \frac{\partial u_z}{\partial y} \right\}$$

(3.16)

$$\tau_{xz} = G_s \left\{ \frac{\partial u_x}{\partial z} + \frac{\partial u_z}{\partial x} \right\}$$

(3.17)

Under the action of a pure normal pressure, $p(\xi, \eta)$, the surface displacements can be determined by setting $z = 0$:

$$u_x = \frac{1}{4\pi G_s}\left\{-(1-2v)\frac{\partial \psi_1}{\partial x} - z\frac{\partial \psi}{\partial x}\right\} = -\frac{1-2v}{4\pi G_s}\left\{\frac{\partial \psi_1}{\partial x}\right\}_{z=0} \tag{3.18a}$$

$$u_y = \frac{1}{4\pi G_s}\left\{-(1-2v)\frac{\partial \psi_1}{\partial y} - z\frac{\partial \psi}{\partial y}\right\} = -\frac{1-2v}{4\pi G_s}\left\{\frac{\partial \psi_1}{\partial y}\right\}_{z=0} \tag{3.18b}$$

$$u_z(x,y) = \frac{1}{4\pi G_s}\left\{2(1-v)\psi - z\frac{\partial \psi}{\partial z}\right\} = \frac{(1-v)}{2\pi G_s}\{\psi\}_{z=0}$$

$$= \frac{(1-v)}{2\pi G_s}\iint_S \frac{p(\xi,\eta)}{R}d\xi d\eta = \frac{(1-v^2)}{\pi E}\iint_S \frac{p(\xi,\eta)}{\sqrt{(x-\xi)^2 + (x-\eta)^2}}d\xi d\eta \tag{3.19}$$

This is the Boussinesq equation for surface normal deformation. Here, the Green's function in Equation (3.19) is

$$G^p = \frac{(1-v^2)}{\pi ER}$$

When two surfaces made of two different materials (E_1, E_2, v_1, v_2) are in contact and subjected to the contact pressure, p, the total normal displacement is the sum of the displacements of each material, as expressed by Equation (3.20):

$$u_{zT} = u_{z1} + u_{z2}$$

$$= \left[\frac{(1-v_1^2)}{\pi E_1} + \frac{(1-v_2^2)}{\pi E_2}\right]\iint_S \frac{p(\xi,\eta)}{\sqrt{(x-\xi)^2 + (y-\eta)^2}}d\xi d\eta = \frac{1}{\pi E'}\iint_S \frac{p(\xi,\eta)}{\sqrt{(x-\xi)^2 + (y-\eta)^2}}d\xi d\eta$$

$$\tag{3.20}$$

The total load is given by the integration of pressure $p(x,y)$ over Ω.

$$W = \iint_S p(x,y)dxdy \tag{3.21}$$

The normal displacement due to a line pressure on a half plane is given in the following Flamant function expressed with x_r as a reference point.

$$u_z(x) = \frac{(1-v^2)}{\pi E}\int_a^b p(\xi)\left[\ln\left|\frac{x-\xi}{x_r-\xi}\right|\right]d\xi = \frac{(1-v^2)}{\pi E}\int_a^b p(\xi)\left[\ln\left|\frac{\xi-x}{\xi-x_r}\right|\right]d\xi \tag{3.22}$$

Friction affects the normal contact pressure if the contacting materials are dissimilar. For contacting materials with the same elastic properties, the additional normal displacements of the contacting bodies, caused by opposite tangential tractions, should be the same in magnitude but opposite in direction. Therefore, friction does not affect the normal Hertz pressure distribution if the materials are not dissimilar. Although the contact equations are still in the domain of linear elasticity, the boundary of Ω is unknown; it is in fact nonlinearly related to pressure, or load. The solution to these equations includes the determination of p, Ω, and z. Efforts of many researchers working on numerical solutions to contact problems involving complex engineering surfaces have yielded rich solutions; among them are the conjugate gradient method (CGM) for rapid pressure boundary search (Polonsky and Keer 2000) and the discrete convolution and fast Fourier transform (DC-FFT) method for accelerated and accurate pressure iteration over the smallest possible computation domain (Liu et al. 2000; Liu and Wang 2002). On the other hand, closed-form formulas for contact pressure, area, and normal approach can be pursued for the contact of ideally smooth surfaces, shown in the following sections.

3.2 SUMMARY OF HERTZ CONTACT THEORIES

3.2.1 ELLIPTICAL AND CIRCULAR CONTACT

Hertz contact theories (Hertz 1881, 1882) solve the contact of smooth surfaces when the contact region is much smaller than the size of the contacting bodies. It is reasonable to assume that the contact area for two ellipsoids is an elliptical region with major and minor radius a and b, so that the following ellipsoidal pressure distribution is true, which yields the maximum pressure, P_h, at the center, and zero pressure outside of the contact area. P_h is called the *maximum Hertzian pressure*.

$$p(x,y) = P_h \sqrt{1 - \frac{x^2}{a^2} - \frac{y^2}{b^2}} \qquad x \leq a, \ y \leq b \qquad (3.23)$$

This problem can be simply described with the contact of an ellipsoid with a half-space material. If u_z is the elastic deformation and z_i the original separation (gap) between the ellipsoid and a reference half-space surface in the contact area, the normal approach of the ellipsoid centers toward the half space is the summation of the elastic deformation and the original surface separations, given in Equation (3.24).

$$\delta = u_z + z_1 + z_2 \qquad (3.24)$$

The corresponding surface normal deformation, $u_z = u_{z1} - u_{z2}$, is (Johnson 1985):

$$u_z = \frac{1}{\pi E^*} \left(L - Mx^2 - Ny^2 \right) \qquad (3.25)$$

Therefore,

$$\frac{1}{\pi E^*}\left(L - Mx^2 - Ny^2\right) = \delta - Ax^2 - By^2 \qquad (3.26)$$

with

$$A = \frac{M}{\pi E^*}, \quad B = \frac{N}{\pi E^*}, \text{ and } \delta = \frac{L}{\pi E^*} \qquad (3.27)$$

Solving elliptical integrations for M, N, and L using $e = (1 - b^2/a^2)^{1/2}$ (with $a \geq b$), A and B, the major and minor contact radii, a and b, can be determined numerically. The contact elliptical ratio can be defined as $K = b/a$. The maximum pressure, P_h, can be solved from integration of pressure over the elliptical area, πab, and balanced with load W. Finally,

$$P_h = \frac{3W}{2\pi ab} \qquad (3.28)$$

Therefore, the pressure distribution is

$$p(x,y) = \frac{3W}{2\pi ab}\sqrt{1 - \frac{x^2}{a^2} - \frac{y^2}{b^2}} \qquad (3.29)$$

Circular contact is a special case of elliptical contact problems, and the orthogonal radii of curvature of each body are the same, $a = b$. Therefore,

$$a = \frac{\pi R_e P_h}{2E^*} \qquad (3.30)$$

$$\delta = \frac{\pi a P_h}{2E^*} = \frac{a^2}{R_e} \qquad (3.31)$$

The stress field caused by the Hertz contact can be illustrated with isograms of the maximum shear stress, τ_{max}, the Tresca stress, given in Equation (3.32), or the von Mises stress, σ_{vm}, given in Equation (3.33), computed for every point under the surface. σ_1, σ_2, and σ_3 in these equations are the principal normal stresses, with σ_1 the maximum and σ_3 the minimum.

$$\tau_{max} = \frac{(\sigma_1 - \sigma_3)}{2} \qquad (3.32)$$

TABLE 3.1

Maximum Tresca Stress and Maximum von Mises Stress along the z Axis

	τ_{max}		τ_{max}		σ_{VM}	
Ellipticity	Johnson (1985) Numerical Solution				Numerical Solution	
$K=b/a$	τ_{max}/P_H	z/a	τ_{max}/P_H	z/a	σ_{VM}/P_H	z/a
0.20	0.322	0.149	0.321	0.149	0.586	0.142
0.40	0.325	0.266	0.325	0.266	0.605	0.260
0.60	0.323	0.354	0.322	0.355	0.615	0.355
0.80	0.317	0.424	0.316	0.425	0.619	0.425
1.00	0.310	0.480	0.310	0.480	0.620	0.480
1.25	0.317	0.530	0.316	0.531	0.619	0.531
1.67	0.323	0.590	0.322	0.592	0.615	0.591
2.50	0.325	0.665	0.325	0.665	0.605	0.650
5.00	0.322	0.745	0.321	0.745	0.586	0.708
Infinity	0.300	0.785	0.302	0.785	0.558	0.703

$$\sigma_{vm} = \sqrt{\frac{(\sigma_1 - \sigma_2)^2 + (\sigma_1 - \sigma_3)^2 + (\sigma_2 - \sigma_3)^2}{2}} \tag{3.33}$$

When friction is absent, the stress distribution is symmetric. The location of maximum von Mises stress increases with increasing K. Table 3.1 lists the numerical results of the locations and values of the maximum Tresca stress and the maximum von Mises stress along the z axis, produced by the Hertz pressure, as well as comparison to Johnson's maximum Tresca stress values (1985), where the data for $b/a < 1$ was obtained through orthogonal conversion of his data for $b/a > 1$ (Wang and Zhu 2011a). The maxima of both the von Mises and Tresca stresses occur at about $z = 0.480a$ for the circular contact of $K = 1$ (Wang and Zhu 2011b). However, locations of the maxima of the two stresses become different when K deviates from 1. The von Mises stresses in the vicinity of its maximum, $z/a \sim 0.68 - 0.72$, are nearly the same when the elliptical ratio is large. Theoretically, for the line contact with $K =$ infinite, the maximum Tresca stress is about $0.300P_h$ at $z = 0.786a$, while the maximum von Mises stress occurs at about $0.558P_h$ at $z = 0.703a$, both calculated from the closed-form line contact stress formulas (Wang and Zhu 2011c).

3.2.2 LINE CONTACT

The line contact of cylinders is a plane-strain problem. If a cylinder of equivalent radius R_e is deformed by a rigid half plane with load P per unit length, the contact area should be a narrow band of width $2a$. Using u_z as the elastic deformation and z the original separation between the cylinder and the half plane surface at $r < a$, the normal approach of the cylinder center toward the half plane is the summation of the elastic deformation and the original surface separation, given by Equation (3.34).

$$\delta = u_z(r) + z(r) \tag{3.34}$$

Here $z(r) \approx r^2/(2R_e)$. The pressure distribution is parabolic, which yields the maximum pressure, $P_h = \dfrac{2P}{\pi a}$, at the center, and zero when $r = a$.

$$p(x,y) = p(x) = P_h\sqrt{1 - \frac{x^2}{a^2}} \quad x \le a \tag{3.35}$$

and

$$a = \sqrt{\frac{4PR_e}{\pi E^*}} \tag{3.36}$$

$$P_h = \left(\frac{PE^*}{\pi R_e}\right)^{1/2} \tag{3.37}$$

The stress components are given in Equation (3.38) (Johnson 1985).

$$\sigma_x = -\frac{2z}{\pi}\int_{-a}^{a} p(\xi)\left[\frac{(x-\xi)^2}{[(x-\xi)^2 + z^2]^2}\right]d\xi - \frac{2}{\pi}\int_{-a}^{a} q(\xi)\left[\frac{(x-\xi)^3}{[(x-\xi)^2 + z^2]^2}\right]d\xi$$

$$\sigma_z = -\frac{2z^3}{\pi}\int_{-a}^{a} p(\xi)\left[\frac{1}{[(x-\xi)^2 + z^2]^2}\right]d\xi - \frac{2z^2}{\pi}\int_{-a}^{a} q(\xi)\left[\frac{x-\xi}{[(x-\xi)^2 + z^2]^2}\right]d\xi \tag{3.38}$$

$$\tau_{xz} = -\frac{2z^2}{\pi}\int_{-a}^{a} p(\xi)\left[\frac{(x-\xi)}{[(x-\xi)^2 + z^2]^2}\right]d\xi - \frac{2z}{\pi}\int_{-a}^{a} q(\xi)\left[\frac{(x-\xi)^2}{[(x-\xi)^2 + z^2]^2}\right]d\xi$$

When the surface shear is neglected, and the stress components along the z axis are the integrations of Equations (3.38) with $x = 0$,

$$\sigma_x = -\frac{P_h}{a}\left\{\frac{(a^2 + 2z^2)}{\sqrt{(a^2 + z^2)}} - 2z\right\} = -P_h\left\{\frac{\left(1 + 2\left(\frac{z}{a}\right)^2\right)}{\sqrt{1 + \left(\frac{z}{a}\right)^2}} - 2\frac{z}{a}\right\} \tag{3.39}$$

$$\sigma_z = -P_h \left\{ \frac{a}{\sqrt{(a^2 + z^2)}} \right\} = -P_h \left\{ \frac{1}{\sqrt{1 + \left(\dfrac{z}{a}\right)^2}} \right\} \qquad (3.40)$$

$$\sigma_y = \upsilon(\sigma_x + \sigma_z) = -\frac{2\upsilon P_h}{a} \left\{ \sqrt{(a^2 + z^2)} - z \right\} = -2\upsilon P_h \left\{ \sqrt{1 + \left(\frac{z}{a}\right)^2} - \frac{z}{a} \right\} \qquad (3.41)$$

σ_x, σ_y, and σ_z given in Equations (3.40) and (3.41) are principal stresses. Here, σ_y is obtained using Hooke's law. σ_z is the smallest among all these stresses; however, the comparison of σ_x, σ_y depends on the z value and Poisson's ratio (Wang and Zhu 2011c). The maximum Tresca stress can be in either $(\sigma_x - \sigma_z)/2$ or $(\sigma_y - \sigma_z)/2$, and the latter is a function of Poisson's ratio. The von Mises stress involves all three principal stresses and is a function of Poisson's ratio. Figure 3.1 shows the values of $(\sigma_x - \sigma_z)/2$ and $(\sigma_y - \sigma_z)/2$ for Poisson's ratio $v = 0.3$ and $v = 0.22$.

For Poisson's ratio $= 0.3$, the maximal Tresca stress is about $0.300P_h$ at $z = 0.785a$, and the maximum von Mises stress is about $0.558P_h$ at $z = 0.703a$. For Poisson's ratio $= 0.22$, the maximal Tresca stress is about $0.316P_h$ at $z = 0.350a$, and the maximum von Mises stress is about $0.596P_h$ at $z = 0.617a$.

The locations of both the maximum Tresca and the von Mises stresses are closer to the surface at a smaller Poisson's ratio. The transition of the maximum Tresca stress from $(\sigma_y - \sigma_z)/2)$ to $(\sigma_x - \sigma_z)/2$ occurs at $v = 0.243$. For $v \geq 0.243$, the maximum appears in $(\sigma_x - \sigma_z)/2$, given below, as a constant, $0.300P_h$ at a fixed location, $z = 0.786a$.

$$\tau_{max} = \frac{P_h}{a} \left\{ z - \frac{z^2}{\sqrt{(a^2 + z^2)}} \right\} = P_h \left\{ \frac{z}{a} - \frac{\left(\dfrac{z}{a}\right)^2}{\sqrt{1 + \left(\dfrac{z}{a}\right)^2}} \right\} \qquad (3.42)$$

3.3 EXTENSION OF THE HERTZ THEORIES TO THE CONTACT OF COATED MATERIALS

Many engineering components have coatings of some kind to modify the surface properties. The displacements and stresses in a layered body can be solved in hybrid space-frequency domains based on the O'Sullivan and King (1988) method extended

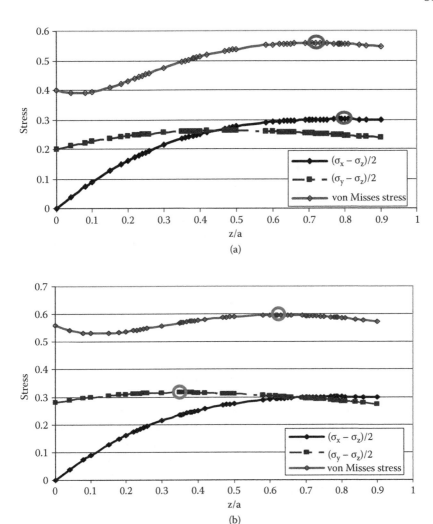

FIGURE 3.1 The von Mises stress and Tresca stress defined by either $(\sigma_x - \sigma_z)/2$ or $(\sigma_y - \sigma_z)/2$ along the z axis. (a) Poisson's ratio $\nu = 0.3$. (b) Poisson's ratio $\nu = 0.22$. The maximum von Mises and Tresca stresses, and their locations (indicated by circles), change with Poisson's ratio.

from the result of Chen (1971). Nogi and Kato (1997) developed explicit frequency response functions of the elastic field for a pressure-loaded layered body. The approaches by O'Sullivan and King, and Nogi and Kato, have been implemented by many researchers to develop numerical solutions for complicated contact problems (Polonsky and Keer 2000; Liu and Wang 2003, 2007; Liu et al. 2007, 2008a). Chen et al. (2010) tackled this problem from a different angle by means of the equivalent inclusion method (Mura 1982).

Figure 3.2 shows several results of the pressure profiles of contacts involving different coating–substrate combinations expressed by the ratios of the Young's

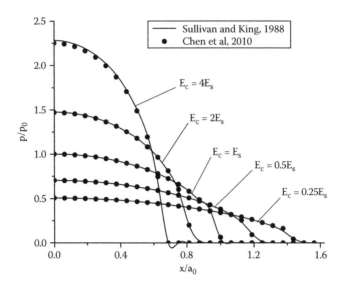

FIGURE 3.2 Contact pressure calculated by Sullivan and King (1988) with the frequency response function method and by Chen et al. (2010) with the equivalence inclusion method.

moduli of the coating and the substrate. Although the curves are different, their overall shapes are alike, and similar to the pressure distribution of the Hertz contact of homogeneous materials. This similarity motivated Liu et al. (2005a, 2005b) to pursue convenient solutions to coating contact problems. The Hertz solutions enjoy explicit formulas for contact radius, normal approach, and the maximum contact pressure. Liu et al. found that the Hertz theories for circular, elliptical, and line contacts can be extended to contact problems through defining a new equivalent Young's modulus and a parameter related to this modulus. The following is the extended Hertz theory by Liu et al. (2005a) for circular contact; the theories for elliptical and line contact problems can be found in the work by Liu et al. (2005a, 2005b).

The frequency response function (FRF) of the normal displacement due to pressure on the surface of a three-dimensional body is given as follows (O'Sullivan and King 1988; Nogi and Kato 1997):

$$\tilde{\tilde{u}} = \frac{1-\nu_c}{\mu_c\, w}\, \frac{1+4wh\kappa\vartheta - \lambda\kappa\vartheta^2}{1-(\lambda+\kappa+4\kappa w^2 h^2)\vartheta + \lambda\kappa\vartheta^2} \tag{3.43}$$

where $\lambda = 1 - \dfrac{4(1-\nu_c)}{1+\mu(3-4\nu_s)}$, $\kappa = \dfrac{\mu-1}{\mu+(3-4\nu_c)}$, $\vartheta = \exp(-2w\,h)$, and $\mu = \mu_c/\mu_s$.

The double tilde means two-dimensional Fourier transform with respect to x and y, and w is defined in the frequency domain and is the counterpart of the radius in the space domain. μ is the shear modulus ratio. The two-dimensional FRF for body 1 is simply the three-dimensional FRF with w replaced by ω, the counterpart of x,

$$\tilde{u} = \frac{1-v_c}{\mu_c\,\omega}\,\frac{1+4\omega h\kappa\theta-\lambda\kappa\theta^2}{1-(\lambda+\kappa+4\kappa\omega^2 h^2)\theta+\lambda\kappa\theta^2} \tag{3.44}$$

where $\theta = \exp(-2\omega h)$ and the tilde means one-dimensional Fourier transform with respect to x. If H is zero (i.e., no coating), the Equation (3.44) becomes the frequency response function for a homogeneous material, as shown in Equation (3.45) with $\dfrac{1-\lambda\kappa}{1-(\lambda+\kappa)+\lambda\kappa}=\dfrac{E_c^*}{E_s^*}$, and $E_i^* = \dfrac{E_i}{1-v_i^2}$.

$$\tilde{u} = \frac{1-v_s}{\mu_s\,\omega} \text{ or } \tilde{u} = \frac{2}{E_s^*\omega} \tag{3.45}$$

Because the coating has different material properties than the substrate, the mechanical responses of a coated half space are due to the combined contribution of the coating and substrate. These responses can be regarded as another equivalent half plane with some equivalent properties. The comparison indicates a new equivalent modulus defined as follows:

$$E_1^* = E_c^*\,\frac{1-(\lambda+\kappa+4\kappa\omega^2 h^2)\theta+\lambda\kappa\theta^2}{1+4\omega\,h\kappa\theta\,-\lambda\kappa\theta^2} \tag{3.46}$$

The Hertzian solutions between body 1 with the substrate alone and body 2 are denoted by a_{0s} and p_{0s}. On the other hand, if the coating is infinitely thick, the corresponding Hertzian solutions are denoted by a_{0c} and p_{0c}. Nondimensional coating thickness is defined as

$$H = h\,/\,a_{0s} \tag{3.47}$$

for elastic substrates, and

$$H = h\,/\,a_{0c} \tag{3.48}$$

for rigid substrates. A parameter, $\alpha = \omega\,h\,/\,H$, is defined; thus, the previous equivalent modulus is a function of material properties (μ_c, μ_s, v_c, v_s), coating thickness H, and parameter α,

$$E_1^* = E_c^*\,\frac{1-(\lambda+\kappa+4\kappa\alpha^2 H^2)\exp(-2\alpha H)+\lambda\kappa\exp(-4\alpha H)}{1+4\alpha H\kappa\exp(-2\alpha H)-\lambda\kappa\exp(-4\alpha H)} \tag{3.49}$$

For the contact between an equivalent body, as body 1, and body 2, the equivalent Young's modulus becomes

$$\frac{1}{E^*} = \frac{1}{E_1^*} + \frac{1-v_2^2}{E_2}$$

This treatment can be done similarly for body 2, if it also has a coated surface. The value of α can be obtained by comparing the numerical results with the predictions with the Hertz formulas. Note that α depends on H and E_c^* / E_s^*.

Quantities $a_{0s}, p_{0s},$ and δ_{0s} obtained for the substrate in the Hertz contact are used to nondimensionalize the contact radius (a_{num}), maximum contact pressure (p_{num}), and contact approach (δ_{num}) determined with the simulation code, $\bar{a} = a_{num} / a_{0s}$, $\bar{p} = p_{num} / p_{0s}$, and $\bar{\delta} = \delta_{num} / \delta_{0s}$. The nondimensional coating thickness is defined with a_{0s} as well. Six curves, shown in Figure 3.3, are selected with modulus ratio of 1/3 and 3, denoted as $\alpha_a(1/3)$ and $\alpha_a(3)$ for contact radius; $\alpha_p(1/3)$ and $\alpha_p(3)$ for maximum contact pressure; and α_δ (1/3) and α_δ (3) for contact approach. Values of α for other different modulus ratios are expressed with modulus ratio E as follows.

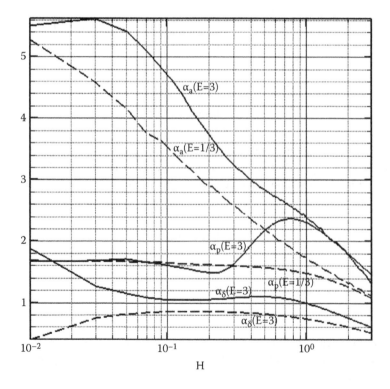

FIGURE 3.3 Representative curves of α for a modulus ratio of 1/3 and 3, denoted as $_a(1/3)$ and $_a(3)$ for contact radius; $_p(1/3)$ and $_p(3)$ for maximum contact pressure; and $_\delta$ (1/3) and $_\delta$ (3) for contact approach.

$$\alpha_p(E) = \begin{cases} \alpha_p(1/3) - (1/E - 3)/14 & E \in [1/4, 1/3] \\ 1.85 + [\alpha_p(1/3) - 1.85]/(3E) & E \in [1/3, 1] \\ 1.72 + E[\alpha_p(3) - 1.72]/3 & E \in [1, 3] \\ \alpha_p(3) - (E - 3)/25 & E \in [3, 4] \end{cases}$$ (3.50)

$$\alpha_\delta(E) = \begin{cases} \alpha_\delta(1/3) - (1/E - 3)/39 & E \in [1/4, 1/3] \\ 0.685 + [\alpha_\delta(1/3) - 0.685]/(3E) & E \in [1/3, 1] \end{cases}$$ (3.51)

With the values of α and E_1^*, Equations (3.28) through (3.31) can be used for contact analysis for $a = b$ (circular contact). The error to quantify the accuracy of the calculation is defined as the difference in percentage between the values predicted with the Hertz formulas and that from numerical solutions. It is found that for all cases analyzed, the errors are less than 1.5%. In real applications, Poisson's ratio may differ from 0.3. The analyses of 2424 cases show that for coating Poisson's ratio varying from 0.15 to 0.3 and modulus ratios selected to be 4, 1.25, 0.8, 0.25, the largest error is still less than 3%.

3.4 SUMMARY OF ADVANCEMENTS ON LUBRICATION MODELING

3.4.1 FUNDAMENTALS OF ELASTOHYDRODYNAMIC LUBRICATION

The Reynolds equation (Equation [3.52]) is the essential equation relating the lubricant film thickness and hydrodynamic pressure. It describes the characteristics of fluid flow through a small clearance bounded by two solid surfaces. One makes the following assumption in deriving the Reynolds equation: the fluid is Newtonian and the flows are laminar; there is no interfacial slippage; the body force is negligible; the lubricant film is thin, and the derivative with respect to the film thickness direction is more important than others; the global curvature effect is insignificant, although surface roughness cannot be neglected; and the pressure across the fluid film can be treated as constant.

$$\frac{\partial}{\partial x}\left(\frac{h^3}{\eta}\frac{\partial p}{\partial x}\right) + \frac{\partial}{\partial y}\left(\frac{h^3}{\eta}\frac{\partial p}{\partial y}\right) = 6U\frac{\partial h}{\partial x} + 12\frac{\partial h}{\partial t}$$ (3.52)

Here, h is the lubricant film thickness between two surfaces, p the hydrodynamic pressure, U the sum of speeds of the two surfaces, and η the viscosity of the lubricant. The coordinates, x and y, are along and perpendicular directions of the motion. This equation describes three types of flows involved in hydrodynamic lubrication: pressure-driven flows, shown by the terms on the left side of the equation, and shear and squeezing flows, shown by the first and second terms on the right side of the equation.

Surfaces under pressure should experience deformation. Lubrication with contribution from surface elasticity is in the category of elastohydrodynamic lubrication, or EHL. EHL is especially the case for components under lubricated counterformal contact. Therefore, the Reynolds equation should be solved together with an elasticity equation.

Both pressure and temperature affect lubricant viscosity. There are many relationships for viscosity influenced by environment. A commonly used exponential viscosity–pressure relationship is given as Equation (3.53) with η_0 the viscosity at the ambient and α the pressure-viscosity coefficient.

$$\eta = \eta_0 e^{\alpha p} \tag{3.53}$$

Bair et al. (2006) suggested that a piezoviscous parameter together with the ambient viscosity would quantify the Newtonian rheology so that the film thickness may be calculated accurately. Several studies have been conducted using viscosity measured in pressure viscometers. Calculated isothermal EHL film thickness values appear to agree with the experiment (Bair et al. 2006; Liu et al. 2008b).

The density of a lubricant may also vary with pressure. The following equation is the density–pressure relationship for mineral oils developed by Dowson and Higgison (1966).

$$\frac{\rho}{\rho_0} = 1 + \frac{0.6p}{1+1.7p} \tag{3.54}$$

Numerically solving these equations together with elasticity should yield the pressure and film thickness at an EHL interface. For a typical EHL problem, the central part of the film is largely flat. A film neck may appear close to the exit portion of the EHL interface, and a pressure spike should accompany the film necking phenomenon.

The multigrid method implemented by Lubrecht (1987), Venner (1991), and Ai (1993), the DC-FFT algorithm composed by Liu et al. (2000), Liu and Wang (2002), the coupled deformation differentiation scheme derived by Holms (2002), and the progressive mesh densification approach proposed by Zhu (2007) and Liu et al. (2009) have been used to improve EHL solution convergence, accuracy, or computational speed. Hu and Zhu (2000) developed a unified solution approach based on the semi-system method (Venner 1991, Ai 1993). With this approach, the solution convergence is further improved for cases under severe conditions. Using this robust approach, one can tackle problems of machined or textured component surfaces in a wide range of operating conditions and with film thickness from several microns down to practically zero.

3.4.2 ROUGH SURFACE EFFECT ON LUBRICATION AND MIXED EHL

Surface microtopography affects lubrication when the film thickness is comparable to the scale of surface roughness (Patir and Cheng 1978). Venner and Lubrecht (1994) computed a single transverse ridge going through the EHL conjunction using the integrated multilevel method. Holmes at al. (2002a, 2002b) investigated start-stop experimental observations by using the coupled differential deflection method. Zhao and Sadeghi (2003) and Zhao (2005) simulated these two transient EHL experiments using a mixed lubrication model that handles the contact area and lubrication area separately. Felix-Quinonez et al. (2003, 2004, 2005) investigated roughness effects of a single ridge and distributive three-dimensional flat-top defects passing through EHL conjunction experimentally and numerically. The Hu-Zhu model (2000) and its recent improvement (Liu et al. 2006a, 2006b, 2009; Liu et al. 2007; Liu et al. 2008a, 2008b,) supported the investigations of a wide range of mixed EHL problems (Wang et al. 2008).

3.4.3 THREE-DIMENSIONAL LINE-CONTACT EHL ON ROUGH SURFACES

Line contacts are found between many mechanical components, such as various gears, roller and needle bearings, cams and followers, and work rolls and backup rolls in metal-forming equipment. The macro aspects of a line-contact problem can be simplified into a two-dimensional model; however, the topography of contacting rough surfaces, micro contacts, and lubricant flows around asperities are often three-dimensional. Ren et al. (2009) developed a three-dimensional line-contact EHL model, shown in Figure 3.4, for solving rough surface lubrication problems.

The elastic deformation of rough surfaces in counterformal contacts can be, in general, analyzed by the DC-FFT algorithm presented by Liu et al. (2000) and Liu

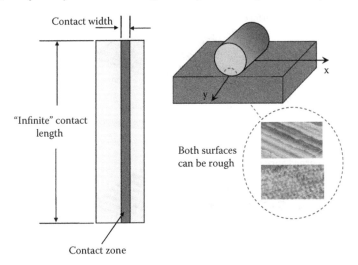

FIGURE 3.4 Two bodies with rough surfaces in a line contact. The contact length in the y direction is infinite, or much greater than the contact width in the x direction.

and Wang (2002), with which the computation domain is formed by extending the physical domain twice in each dimension. Excitation in the extended domain is padded with zero so as to circumvent the otherwise border-aliasing error. In order to solve the three-dimensional line-contact problem, the DC-FFT algorithm has been advanced by Chen et al. (2008a) with the mixed padding strategy in the extended domain. For line contact problems, the contact zone is infinitely long in one direction but finite in the other. The surface roughness along the direction of the infinite length is assumed to be periodically repeated. The mixed padding method can then be used to duplicate padding excitation in the direction of the infinitely long contact geometry and zero padding excitation in the direction of the finite contact width. The influence coefficient matrix is formed by the wrap-around order approach similar to that in the DC-FFT algorithm. With this approach, deformation can be computed in the chosen domain of limited length and width, although it is actually caused by the pressure distributed not only inside but also outside of the computation domain.

To illustrate, consider a honed surface, with root-mean-square roughness $R_q = 128$ nm with an obvious texture, mating with a turned surface of larger roughness ($R_q = 391$ nm), as shown in Figure 3.5. In this case, the dimensionless parameters are $G = 4467.8$, $W = 0.5308 \times 10^{-3}$, $U = 0.2551 \times 10^{-10}$, and the sliding-to-rolling ratio $S = 0.25$. The maximum Hertzian pressure is 2.121 GPa, and the half width of the Hertzian zone is 0.294 mm. Solution results for film thickness and pressure distributions are shown in Figure 3.6 (Wang et al. 2008). It was found that the transverse turned surface yields an average film thickness of 250.7 nm, while the longitudinally oriented one gives the film a thickness of 197.8 nm under the same conditions. Because the corresponding smooth-surface EHL film thickness is about 230 nm, it appears that the transverse roughness enhances the film thickness in such a line contact while the longitudinal roughness may cause a film thickness

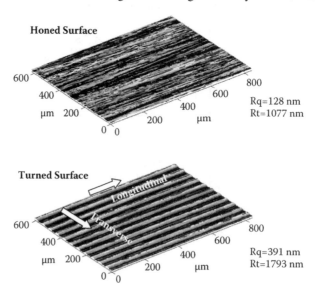

FIGURE 3.5 A honed and a turned surface.

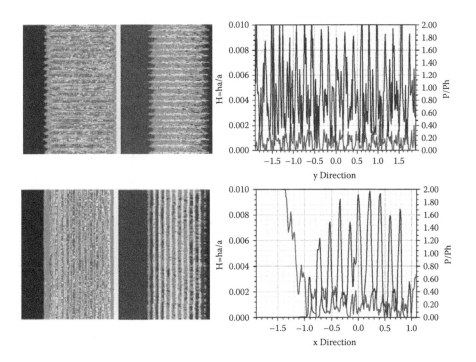

FIGURE 3.6 Lubrication film thickness and pressure distribution plots for the contact between honed and turned surfaces in longitudinal and transverse orientations. Top: Solution for the honed surface vs. the turned surface, longitudinal (film thickness, pressure, cross section of pressure and film thickness along the cylinder length direction normalized by contact half width); Bottom: Solution for the honed surface vs. the turned surface, transverse (film thickness, pressure, cross section of pressure and film thickness perpendicular to the cylinder length direction normalized by contact half width).

reduction. This result is qualitatively in agreement with the results from the stochastic model presented by Patir and Cheng (1978). Quantitatively, the stochastic model appears to yield a greater influence than that shown in the deterministic approach (Ren et al. 2009).

3.4.4 EHL of Elements with Coated Materials

Figure 3.7 shows the EHL contact of coated surfaces modeled by Liu et al. (2007, 2008a). The key to modeling the EHL of a coated surface is the elasticity of layered materials mentioned in Section 3.3. For a coated half-space surface, an explicit Green's function is not available for the normal displacement in the space domain; instead, the frequency response function can be used. After applying the inverse Fourier transfer to the discrete matrix of the frequency response function, influence coefficients can be picked up from the transformed matrix. The elastic deformation of coated surfaces in EHL can then be evaluated by the DC-FFT fast algorithm (Liu et al 2000).

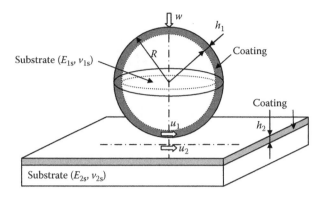

FIGURE 3.7 EHL for coated surfaces in a point contact.

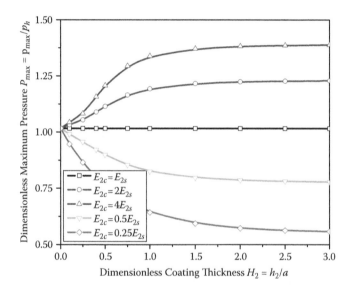

FIGURE 3.8 Effect of coating on maximum contact pressure.

Figure 3.8 presents results for the lightly loaded case of 100 N under a low speed of 0.2 m/s. A hard coating increases the maximum pressure but reduces the nominal contact radius, while the situation is the opposite for a soft coating. However, the change caused by using a soft coating is larger than that by using a hard coating. As the coating thickness increases, the maximum pressure gradually approaches the limits where the substrate is replaced by the coating material completely. As the coating thickness approaches three times the Hertzian contact radius, the elastic deformation behavior is completely controlled by the coating material (Liu et al 2007).

Two dimensionless parameters M and L, defined in Equations (3.55) and (3.56) and suggested by Moes and Bosma (1971) are employed as independent variables to describe the EHL behavior so that no redundant equalities are introduced into the system.

$$M = \frac{w}{E'R^2}(\frac{2\eta_0 u}{E'R})^{-3/4} \tag{3.55}$$

$$L = \alpha E'(\frac{2\eta_0 u}{E'R})^{1/4} \tag{3.56}$$

If a disk surface is uniformly coated with a different material of h_2 in thickness, two dimensionless coating parameters are employed to describe the coating properties, the dimensionless coating thickness and material parameter ratio of the coating–substrate system, as shown in Equations (3.57) and (3.58).

$$H_2 = \frac{h_{2c}}{a} \tag{3.57}$$

$$R_2 = \frac{E_{2c}^*}{E_{2s}^*} = \frac{1-v_{2s}^2}{1-v_{2c}^2}\frac{E_{2c}}{E_{2s}} \tag{3.58}$$

A parametric study is conducted for stiff coatings for a wide range of dimensionless load parameters M, dimensionless material parameters L, and coating material parameters R_2, as well as coating thickness parameters H_2. Totally, 21 combinations of dimensional and dimensionless parameters are considered for different load, speed, coating thickness, and M and L values. For each condition, dimensionless coating thickness H_2 changes from 0.1 to 1 with 9 interval points (0, 0.1, 0.2, 0.25, 0.3, 0.4, 0.5, 0.75, 1.0). The numerical parameter combinations can be summarized into four typical cases, or curves, expressed in Figure 3.9. The four cases correspond to four extreme working conditions respectively for heavy load, light load, low pressure-viscosity, and high pressure-viscosity exponents. In this figure, the horizontal axis is the dimensionless coating thickness, H_2, while the vertical axis refers to the minimum film thickness from the coated case divided by that from the corresponding uncoated case. A ratio of I_{max} larger than 1 indicates improvement on film thickness. When the coating thickness approaches three times the Hertzian radius obtained from the non-coating contact, deformation and film thickness are almost completely controlled by the coating material. For heavy load (larger M), or high pressure-viscosity exponent (larger L), the minimum film thicknesses at $H_2 = 3$ are larger than that for corresponding uncoated cases. However, for the light load (smaller M) case, or the case with low pressure-viscosity exponent (smaller L), the film thickness values are smaller than those for uncoated cases.

Figure 3.9 further confirms that the minimum film thickness improvement using a stiff coating is greatest when the contact is well inside the elastic-piezoviscous

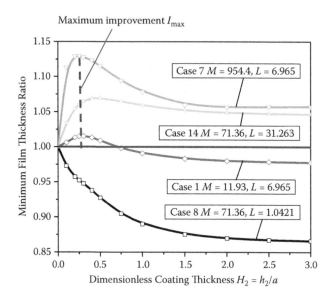

FIGURE 3.9 Variation of minimum film thickness with coating thickness for four extreme working conditions.

regime. The film thickness improvement in the elastic-piezoviscous regime may be a compromised result of two effects. Decreasing elastic deformation due to the increasing surface Young's modulus tends to reduce the film thickness, while increasing normal approach due to the increasing pressure–viscosity effect tends to increase the film thickness. For the case with a thin stiff coating, the latter effect might exceed the former at two side-lobe areas. Therefore, film thickness improvement could be achieved (Liu et al. 2008a).

Based on the numerical simulation results, the maximum increase of minimum film thickness due to coating, expressed with I_{max} and the corresponding optimal dimensionless coating thickness, H_{2max}, are correlated to M, L, and R_2 through multiple regressions. Two curve-fitting formulas are reported as follows:

$$I_{max} = 0.769 M^{0.0238} R_2^{0.0297} L^{0.1376} \exp(-0.0243 \ln^2 L) \qquad (3.59)$$

$$H_{2max} = 0.049 M^{0.4557} R_2^{-0.1722} L^{0.7611} \exp(-0.0504 \ln^2 M - 0.0921 \ln^2 L) \qquad (3.60)$$

These formulas can be used to estimate possible EHL film thickness increases due to the use of a stiff coating in various EHL conditions (Liu et al. 2008a).

3.4.5 Plasto-EHL (PEHL)

Once the stress reaches the initial yield limit of a contacting material, subsurface and surface plastic deformation may happen. The material property may also be altered if the material exhibits work-hardening. These changes can permanently alter the surface profiles and contact geometry, as well as the subsurface stress field even after the applied load is removed. The plastically modified contacting surfaces will in turn generate renewed hydrodynamic pressure and film thickness profiles, sometimes significantly different from those of the initial EHL solution. A full-scale plasto-EHL (PEHL) model has been developed by Ren et al. (2010), which is the extension of the deterministic mixed EHL model originally presented by Hu and Zhu (2000), and modified by Liu et al. (2006a, 2009), and Zhu (2007). The basic mathematic equations describing the PEHL model include those for the EHL formulation together with material plasticity based on the elastic-plastic contact studies by Jacq et al. (2002), Nelias et al. (2007), Chen and Wang (2008), and Chen et al. (2008a and 2008b). Detailed formulation and derivation can be found in the recent work by Ren (2009).

A typical example of a PEHL solution, in comparison with its corresponding EHL solution, is given in Figure 3.10, demonstrating basic PEHL characteristics of a smooth-surface point contact. In this case, a rigid ball of 9.525 mm in diameter is in contact against a stationary elastic perfectly plastic flat with yield strength of 500 MPa, similar to the case shown in Figure 3.4. The rigid ball moves at a speed of U_1 = 0.5m/s and the applied load is 30 N, considerably greater than the critical load W_y. The lubricant is assumed to be Newtonian with dynamic viscosity of 0.0112 Pa.S. Figure 3.10 shows three-dimensional EHL and PEHL pressure distributions together with two-dimensional pressure and film thickness profiles along the moving direction at $y = 0$. The figure indicates that the overall pressure and film thickness profiles of the PEHL solution still somewhat resemble those of the EHL solution, showing some typical characteristics, such as pressure distribution possibly with a pressure spike on the outlet side, reasonably constant film thickness in the central part of the interaction zone, and a film constriction downstream right after the pressure spike. However, the PEHL pressure distribution has a much flatter top portion with a reduced maximum value, while the central film thickness may also be considerably smaller in comparison with those in the EHL solution. In addition, the contact zone is slightly expanded due to the plastic deformation. These observations are consistent with those reported in an elastic-plastic dry contact study by Wang and Keer (2005), showing that plasticity can be effectively integrated into lubrication modeling. Generally, the EHL model predicts a higher pressure and thicker lubricant film. The differences become more pronounced with increasing load.

These results also indicate that stronger work-hardening produces relatively higher surface pressure and thicker lubricant film under the same operating conditions, while weaker work-hardening causes more surface plastic deformation and a more expanded contact area to provide extra load support. The PEHL solution further approaches that of the EHL when the work-hardening gets stronger.

3.5 CHALLENGES AND OPPORTUNITIES IN CONTACT AND LUBRICATION RESEARCH

When the lubricant film thickness approaches the nanoscale, the interface formed by surfaces and the lubricant becomes a complex physical and chemical system, where

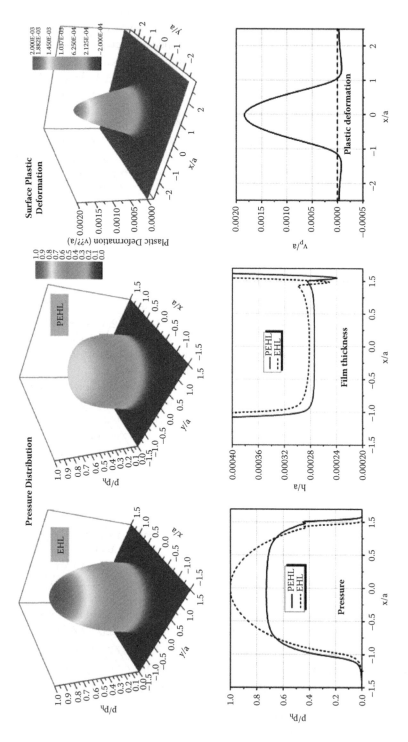

FIGURE 3.10 A typical PEHL solution in comparison with its corresponding EHL solution.

many phenomena are waiting to be discovered. Many exploratory studies by means of molecular dynamics simulations investigated fluid–surface interactions at ideal interface configurations (Hu et al., 1998; Martini et al. 2006, 2008). Viscosity, solvation pressure, and density varying with the film thickness (Martini et al. 2006); wall slip (Martini et al. 2008), and lubricant molecule ordering (Zhang et al. 2004) may all exist together with wear, lubricant degradation, and localized thermodynamic phenomena. Interactive mechanical-physical-chemical studies are needed to unfold the nature of lubrication phenomena at the nanoscale.

Contact failure, fatigue life, and performance prediction. The goal of many tribological studies is to explore contact fatigue and to develop failure prediction methodologies. However, contact fatigue may involve physical, material, and mechanical processes related to the contact interface, such as localized surface tractions and damage accumulation. Although great efforts have yielded approaches toward assessments of contact fatigue failures (Zaretsky 1987; Epstein et al. 2003; Holmes et al. 2004; Qiao et al. 2008; Zhu et al. 2009), more accurate and realistic models for contact performance, failure transition, and life prediction require integrated modeling systems for complex and transient tribological interfaces.

Extreme-condition tribology. Advancements of engineering and technology are challenging the design of tribological systems subjected to extreme conditions, such as extremely small size, extremely low or high temperatures, high stresses, high vacuum, high speed, high contamination, and high life cycle. On the other hand, more and more new materials, lubricants, and additives, and new surface textures have been designed. Application of these new technologies to solving extreme-condition tribological problems requires exploration of fundamental principles to support new designs and consider their interactions.

Surface texturing and surface engineering. The bottleneck of surface engineering is the understanding of complexity of surfaces in tribological interfaces. Surface texture shape, size, depth, density, and patterns of distribution (Wang and Zhu 2005; Zhu et al. 2010), materials of the surface layer containing the topography (Bhushan 1995), contact type and status of surface interactions (Wang et al. 2008), operating conditions and their transitions (Kovalchenko et al. 2004), surface interaction with lubricants, and wear are among the factors of the complication. Developing durable, cost-effective surfaces for tribological applications is the grand challenge of engineering tribology, requiring integration of efforts in mechanical engineering, manufacturing, materials and surface science, and chemistry.

3.6 SUMMARY

This chapter presents the fundamentals of contact and lubrication, together with the review of several advancements of lubrication theories and modeling efforts that combine advanced contact and lubrication mechanics, such as mixed-lubrication rough surfaces, three-dimensional line-contact elastohydrodynamic lubrication, lubrication of coated surfaces, and plasto-elastohydrodynamic lubrication of engineering surfaces. Challenges in contact and lubrication are briefly discussed, such as issues related to lubrication at the nanoscale, extreme-condition tribology, and contact interfacial failure and prediction, as well as durable and cost-effective engineering of tribological interfaces.

REFERENCES

Ai, X. 1993. "Numerical Analysis of Elastohydrodynamically Lubricated Line and Point Contacts with Rough Surfaces by Using Semi-System and Multigrid Methods." PhD thesis, Northwestern University, Evanston, IL.

Bair, S., Liu, Y., and Wang, Q. 2006. "The Pressure-Viscosity Coefficient for Newtonian EHL Film Thickness with General Piezoviscous Response. ASME *Journal of Tribology*, 128: 624–631.

Bhushan, B. 1995. *Handbook of Micro/Nano Tribology*. Boca Raton, FL: CRC Press.

Chen, W.T. 1971. "Computation of Stresses and Displacements in a Layered Elastic Medium." *Int. J. Eng. Sci.* 9, 775–799.

Chen, W.W., Liu, S.B., and Wang, Q.J. 2008a. "Fast Fourier Transform–Based Numerical Methods for Elasto-Plastic Contacts of Nominally Flat Surfaces." *ASME J. Appl. Mech.* 75: 011022.

Chen, W.W., Wang, Q., Wang, F., Keer, L.M., and and Cao, J. 2008b. "Three-Dimensional Repeated Elasto-Plastic Point Contact, Rolling and Siding." *Journal of Applied Mechanics* 75: 021021-1-12.

Chen, W.W., and Wang, Q. 2008. "A Numerical Model for the Point Contact of Dissimilar Materials Considering Tangential Tractions." *Mechanics of Materials* 40: 936–948.

Dowson, D., and Higginson, G.R., 1996. *Elastohydro-Dynamic Lubrication*. Oxford, UK: Pergamon Press.

Epstein, D., Yu, T., Wang, Q., Keer, L.M., Cheng, H.S., Harris, S., and Gangopadhyay, A. 2003. "An Efficient Method of Analyzing the Effect of Roughness on Fatigue Life in Mixed-EHL Contact." *Tribology Transactions* 46: 273–281.

Felix-Quinonez, A., Ehret, P., and Summers, J.L. 2003. "New Experimental Results of a Single Ridge Passing through an EHL Conjunction." *ASME Journal of Tribology* 125: 252–259.

Felix-Quinonez, A., Ehret, P., and Summers, J.L. 2004. "Numerical Analysis of Experimental Observations of a Single Transverse Ridge Passing through an Elastohydrodynamic Lubrication Point Contact under Rolling/Sliding Conditions." *Proc. Instn. Mech. Engrs.*, *Part J: J. of Engineering Tribology* 218: 109–123.

Felix-Quinonez, A., Ehret, P., and Summers, J.L. 2005. "On Three-Dimensional Flat-Top Defects Passing through an EHL Point Contact: A Comparison of Modeling with Experiments." *ASME Journal of Tribology* 127: 51–59.

Hertz, H., 1881/1896. "On the Contact of Elastic Solids." In *Miscellaneous Papers*. New York: MacMillan and Co., Ltd.

Hertz, H., 1882/1896. "On the Contact of Rigid Elastic Solids and on Hardness." In *Miscellaneous Papers*. New York: MacMillan and Co., Ltd.

Hills, D. A., Nowell, D., and Sackfield, A. 1993. *Mechanics of Elastic Contact*. Oxford, UK: Butterworth-Heinemann.

Holmes, M.J.A. 2002. "Transient Analysis of the Point Contact Elastohydrodynamic Lubrication Problem Using Coupled Solution Methods." PhD thesis, Cardiff University, UK.

Holmes, M.J.A., Evans, H.P., and Snidle, R.W. 2002a. "Comparison of Transient EHL Calculations with Start-up Experiments." *Proceedings of the 29th Leeds-Lyon Symposium on Tribology*, D. Dowson et al., eds., 79–89. Amsterdam: Elsevier.

Holmes, M.J.A., Evans, H.P., and Snidle, R.W. 2002b. "Comparison of Transient EHL Calculations with Shut-down Experiments." *Proceedings of the 29th Leeds-Lyon Symposium on Tribology*, D. Dowson et al., eds., 91–99. Amsterdam: Elsevier.

Holmes, M.J.A., Qiao, H., Evans, H.P., and Snidle, R.W. 2004. "Surface Contact and Damage in Micro-EHL." In *Life Cycle Tribology*, *Proc. 31st Leeds-Lyon Symposium on Tribology*, Leeds, Tribology and Interface Engineering Series, 605–616. Amsterdam: Elsevier.

Hu, Y.Z., Wang, H., Gao, Y., and Zheng, L.Q. 1998. "Molecular Dynamics Simulation of Ultra-Thin Lubricating Films. *Proc Instn Mech Engrs* 212, Part J: 165.

Hu, Y.Z., and Zhu, D. 2000. "A Full Numerical Solution to the Mixed Lubrication in Point Contacts." *ASME J. Tribol.* 122(1): 1–9.

Jacq, C., Nelias, D., Lormand, G., and Girodin, D. 2002. "Development of a Three-Dimensional Semi-Analytical Elastic-Plastic Contact Code." *ASME J. Tribol.* 124: 653–667.

Johnson, K. L. 1985. *Contact Mechanics*. Cambridge, UK: Cambridge University Press.

Kovalchenko, A., Ajayi, O., Erdemir, A., Fenske, G., and Etsion, I. 2004. "The Effect of Laser Texturing of Steel Surfaces and Speed-Load Parameters on the Transition of Lubrication Regime from Boundary to Hydrodynamic." *Tribology Transactions* 47: 299–307.

Liu, G., and Wang, Q. 2000. "Thermoelastic Asperity Contacts, Frictional Shear, and Parameter Correlation." ASME *Journal of Tribology* 122: 300–307.

Liu, S., Wang, Q., and, Liu, G. 2000. "A Versatile Method of Discrete Convolution and FFT (DC-FFT) for Contact Analyses." *Wear* 243: 101–111.

Liu, S.B., and Wang, Q. 2002. "Study Contact Stress Fields Caused by Surface Tractions with a Discrete Convolution and Fast Fourier Transform Algorithm." *ASME J. Tribol.* 124: 36–45.

Liu, S., and Wang, Q. 2003. "Transient Thermoelastic Stress Fields in a Half-Space." *ASME Journal of Tribology* 125: 33–43.

Liu, S., Peyronnel, A., Wang, Q, and Keer, L. 2005a. "An Extension of the Hertz Theory for 2D Coated Components." *Tribology Letters* 18: 505–511.

Liu, S., Peyronnel, A., Wang, Q, and Keer, L. 2005b. "An Extension of the Hertz Theory for Three-Dimensional Coated Bodies." *Tribology Letters*. 18: 303–314.

Liu, Y., Wang, Q., Hu, Y., Wang, W., and Zhu, D. 2006a. "Effects of Differential Schemes and Mesh Density on EHL Film Thickness in Point Contacts." *ASME Journal of Tribology* 128: 641–653.

Liu, Y.C., Wang, Q.J., Wang, W.Z., Hu, Y.Z., Zhu, D., Krupka, I. and Hartl, M., 2006b. "EHL Simulation Using the Free-volume Viscosity Model." *Tribology Letters* 23(1): 27–37.

Liu, S., and Wang, Q. 2007. "Determination of Young's Modulus and Poisson's Ratio for Coatings." *Surface and Coatings Technology* 201: 6470–6477.

Liu, Y., Chen, W., Liu, S., Zhu, D., and Wang, Q. 2007. "An Elastohydrodynamic Lubrication Model for Coated Surfaces in Point Contacts." *Journal of Tribology* 129: 509–516.

Liu, Y., Zhu, D., and Wang, Q. 2008a. "Effect of Stiff Coatings on EHL Film Thickness in Point Contacts." *Journal of Tribology* 130: 031501-1-6.

Liu, Y., Wang, Q., Krupka, I, Hartl, M, and Bair, S. 2008b. "The Shear-Thinning Elastohydrodynamic Film Thickness of a Two-Component Mixture." *Journal of Tribology* 130: 021502-1-7.

Liu, Y., Wang, Q., Zhu, D., Wang, W., and Hu, Y. 2009. "Effects of Differential Scheme and Viscosity Model on Rough-Surface Point-Contact Isothermal EHL." *Journal of Tribology* 131(4): 044501-1-5, 10.1115/1.2842245.

Love, A.E.H. 1929. "Stress Produced in a Semi-Infinite Solid by Pressure on Part of the Boundary." *Phil. Trans. Royal Society* A228: 337.

Love, A.E.H. 1952. *A Treatise on the Mathematical Theory of Elasticity*, 4th ed. Cambridge, UK: Cambridge University Press.

Lubrecht, A.A. 1987. "The Numerical Solution of Elastohydrodynamic Lubricated Line and Point Contact Problems Using Multigrid Techniques." PhD thesis, University of Twente, The Netherlands.

Martini, A., Liu, Y., Snurr, R., and Wang, Q. 2006. "Molecular Dynamics Characterization of Thin Film Viscosity for EHL Simulation." *Tribology Letters* 21: 217–225.

Martini, A., Roxin, A., Snurr, R. Q., Wang, Q., and Lichter, S. 2008. "Molecular Mechanisms of Liquid Slip." *Journal of Fluid Mechanics* 600: 257–269.

Moes, H., and Bosma, R. 1971. "Design Charts for Optimum Bearing Configuration, Part 1, The Full Journal Bearing." *ASME J. Lubr. Tech.* 93: 302–306.

Mura, T., 1982. *Micromechanics of Defects in Solids*. Boston: Martinus Nijhoff.

Nelias, D., Antaluca, E., and Boucly, V. 2007. "Semianalytical Model for Elastic-Plastic Sliding Contacts." *ASME J. Tribol.* 129: 761–771.

Nogi, T., and Kato, T. 1997. "Influence of a Hard Surface Layer on the Limit of Elastic Contact-Part I: Analysis Using a Real Surface Model." *ASME Journal of Tribology* 119: 493–500.

O'Sullivan, T.C., and King, R.B. 1988. "Sliding Contact Stress Field Due to a Spherical Indenter on a Layered Elastic Medium." *ASME Journal of Tribology* 110: 235–240.

Patir, N., and Cheng, H.S. 1978. "An Average Flow Model for Determine Effects of Three Dimensional Roughness on Partial Hydrodynamic Lubrication." ASME *Journal of Lubrication Technology* 100: 12–17.

Polonsky, I. A., and Keer, L.M. 2000. "A Fast and Accurate Method for Numerical Analysis of Elastic Layered Contacts." *ASME Journal of Tribology* 122: 30–35.

Qiao, H., Evans, H.P., and Snidle, R.W. 2008. "Comparison of Fatigue Model Results for Rough Surface Elastohydrodynamic Lubrication." *Proc. IMechE Part J: J Engineering Tribology* 222: 381–393.

Ren, N. 2009. "Advanced Modeling of Mixed Lubrication and Its Mechanical and Biomedical Applications." PhD thesis, Northwestern University, Evanston, IL.

Ren, N., Zhu, D., Chen, W., W., Liu, Y., and Wang, Q. 2009. "A Three-Dimensional Deterministic Model for Rough Surface Line-Contact EHL." *Journal of Tribology* 131: 011501-1-9.

Ren, N., Zhu, G., Chen, W.W., and Wang, Q. 2010. "Plasto-Elastohydrodynamic Lubrication (PEHL) in Point Contact." *Journal of Tribology* 132: 031501-1, DOI: 10.1115/1.4001813.

Venner, C.H. 1991. "Multilevel Solution of EHL Line and Point Contact Problems." PhD dissertation, University of Twente, The Netherlands.

Venner, C.H., and Lubrecht, A.A. 1994. "Numerical Simulation of a Transverse Ridge in a Circular EHL Contact under Rolling/Sliding." *ASME Journal of Tribology* 116(4): 751–761.

Wang, F., and Keer, L.M. 2005. "Numerical Simulation for Three-Dimensional Elastic-Plastic Contact with Hardening Behavior." *ASME Journal of Tribology* 127: 494–502.

Wang, Q., and Zhu, D. 2005. "Virtual Texturing: Modeling the Performance of Lubricated Contacts of Engineered Surfaces." *ASME Journal of Tribology* 127: 722–728.

Wang, Q., Zhu, D., Zhou, R., and Hashimoto, F. 2008. "Investigating the Effect of Surface Finish and Texture on Mixed EHL of Rolling and Rolling-Sliding Contacts." *Tribology Transactions* 51: 748–761.

Wang, Q., and Zhu, D. 2011a. "Hertz Theory: Contact of Elliptical Surfaces." In *Encyclopedia of Tribology*, submitted, to be published by Springer.

Wang, Q., and Zhu, D. 2011b. "Hertz Theory: Contact of Circular Surfaces." In *Encyclopedia of Tribology*, submitted, to be published by Springer.

Wang, Q., and Zhu, D. 2011c. "Hertz Theory: Contact of Cylindrical Surfaces." In *Encyclopedia of Tribology*, submitted, to be published by Springer.

Zaretsky, E.V. 1987. "Fatigue Criterion to System Design, Life and Reliability." *J. Propulsion and Power* 3(1): 76–83.

Zhang C.H, Luo J.B., and Wen, S.Z. 2004. "Exploring Micropolar Effects in Thin Film Lubrication." *Science in China Series G-Physics Mechanics & Astronomy* 47: 65–71.

Zhao, J., 2005. "Analysis of EHL Circular Contact Start Up: Comparison with Experimental Results." *Proceedings of WTC2005 World Tribology Congress III*, September 12–16, Washington, DC, WTC2005-63705.

Zhao, J., and Sadeghi, F. 2003. "Analysis of EHL Circular Contact Shut Down." *ASME Journal of Tribology* 125: 76–90.

Zhu, D. 2007. "On Some Aspects in Numerical Solution of Thin-Film and Mixed EHL." *Proc. IMech, Part J, Journal of Engineering Tribology* 221: 561–579.

Zhu, D., Nanbu, N., Ren, N., Yasuda, Y., and Wang, Q. 2010. "Model-Based Virtual Surface Texturing for Concentrated Conformal-Contact Lubrication." *Proceedings of the Institution of Mechanical Engineers, Part J, Journal of Engineering Tribology,* vol. 224: 686–696.

Zhu, D., Ren, N., and Wang, Q. 2009. "Pitting Life Prediction Based on 3-D Line Contact Mixed EHL Analysis and Subsurface von Mises Stress Calculation." *Journal of Tribology* 131: 041501-1.

4 Surface Energy and Surface Forces

Robert W. Carpick

CONTENTS

4.1 INTRODUCTION

The large surface-to-volume ratio intrinsic to nanoscale materials, structures, and devices ensures that surface forces are dominant. The dominance of surface forces at small scales is present wherever we look. Dust, insects, and other small entities can stick to walls and ceilings, while large objects obviously do not. Forces often thought of as weak, including van der Waals, electrostatic, and capillary interactions, are sufficient to overcome gravity at small scales. Other bulk-related or mass-determined forces, including inertial effects, are often not enough to displace such small entities, and this has proven to be a critical issue in small devices. For example, the Digital Micromirror Device™ developed by Texas Instruments required significant effort to overcome adhesion problems [1] (Figure 4.1). The device, which is widely used for commercial digital image projection equipment, consists of an array of microfabricated mirrors, each on a flexible hinge. Each mirror corresponds to one pixel of the image to be projected. Electrostatic forces can deflect each mirror individually to different angles to reflect the chosen color of light for the pixel. The stopping position of the mirror is crucial, and so its yoke is stopped by making contact with a landing area. As originally fabricated, the designers found the yoke and pad stuck together and could not be released reliably. Coating the surfaces with an organic film, reducing the contact time by vibrating the mirror, adding a spring tip that pushes back when the mirror is released, and hermetically sealing it in a clean, dry environment to prevent condensation of water and contaminants all contributed to solving the problem [1], and the device is now a commercial success, sold in the tens of millions.

As another example, the flexible parts of micro-electromechanical systems (MEMS) devices were often found to be stuck together after being fabricated (Figure 4.2). Coating the surfaces with low-energy molecular groups solved the problem [2]. MEMS devices with contacting surfaces are now sufficiently reliable for commercialization. However, serious issues related to friction and wear still prevent MEMS devices that involve sliding surfaces from being commercialized at this time [3].

The increase in surface versus bulk atoms can be quantified by examining the number of atoms at the surface versus the bulk as a function of particle size. This is illustrated in Figure 4.3 [4] for a face-centered cubic (FCC) nanoparticle; even with over 1400 atoms total in the particle, more than 1/3 of the atoms are at the surface. As another metric, in its typical structure, a 3-nm diameter iron particle has 50% of atoms on the surface, while a 30-nm iron particle has 5% of its atoms on surface.

The high surface-to-volume ratio at the nanoscale has consequences that go far beyond the dominant presence of surface forces. Mechanical behavior due to the confinement or starvation of dislocations and their interactions with surfaces, electrical properties due to quantum confinement and surface-related electronic states, as well as thermal, vibrational, optical, magnetic, and other properties can all display dramatic differences at the nanoscale due to the presence of such a large fraction of surface or interface atoms.

Understanding tribological issues at small scales therefore requires a clear understanding of the basic phenomena related to surface energies and surface forces. This chapter is meant to provide a tutorial overview of basic concepts of surface energy, surface forces, and adhesion, which can then be applied to develop a better understanding of nanotribological phenomena. The chapter begins with the definition of surface energy for liquids and a discussion of how surface tension and capillary pressure originate and are manifested. The Young-Laplace equation, capillary forces, and the Kelvin equation are presented and discussed. The concept of surface energy for solids is then presented. This is followed by a discussion of interfacial energy and its relation to cohesion and adhesion. Techniques for measuring surface and interfacial energy, including contact angle measurements, will be summarized. Surface forces, beginning with a description of the origins of surface forces, are then discussed. The relation between surface energy and surface forces is discussed with reference to the Derjaguin approximation and contact mechanics. Finally, the role that surface energy and surface forces play at small length scales for applications are highlighted

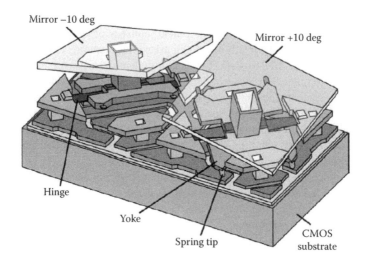

FIGURE 4.1 Schematic of the Digital Micromirror Device (DMD). Permission to reproduce pending from P.F. Vankessel, L.J. Hornbeck, R.E. Meier, and M.R. Douglass, "MEMS-Based Projection Display." *Proc. IEEE* 86(1998): 1687, Figure 1.

FIGURE 4.2 Example of a MEMS test structure. Sufficiently long (i.e., compliant) cantilevers are stuck down to the substrate after the release step. This adhesion phenomenon will render MEMS devices nonfunctional. Permission to reproduce pending from R. Maboudian and R.T. Howe, "Critical Review: Adhesion in Surface Micromechanical Structures." *J. Vac. Sci. Technol.* 15 (1997): 1–20, Figure 15(c).

Full-shell Clusters		Total Number of Atoms	Surface Atoms (%)
1 Shell		13	92
2 Shells		55	76
3 Shells	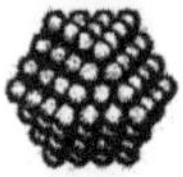	147	63
4 Shells	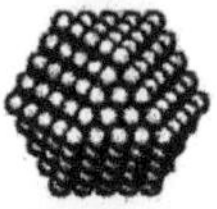	309	52
5 Shells	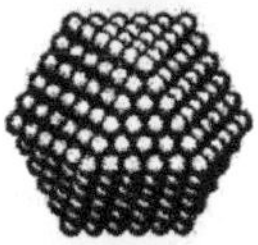	561	45
7 Shells	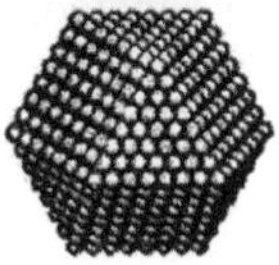	1415	35

FIGURE 4.3 FCC clusters of atoms illustrating how the fraction of surface atoms increases dramatically at the nanoscale. Permission to reproduce pending from K.J. Klabunde and R.M. Richards, eds., *Nanoscale Materials in Chemistry* (Hoboken, NJ: Wiley, 2009), Table 2.1.

with a few selected examples. For a more detailed, in-depth discussion of this topic, the reader is referred to the book by C.M. Mate, "Tribology on the Small Scale" [5].

4.2   SURFACE ENERGY

The term *surface energy* is in fact the source of some confusion, as it can have multiple meanings. First, in all cases where it is used to describe a material property, it is not an energy, but an energy per unit surface area (i.e., with units of J/m^2, for example). The words *per unit area* are often simply dropped for brevity. However, there are times when the absolute surface energy, in Joules for example, is of interest. In such cases, it is important to clearly state that this is the case. Obviously, the absolute surface energy is not an intrinsic material property as it depends on the total amount of surface present for the specific sample of interest.

The terminology is also somewhat confusing, since different terms are generally used for liquids versus solids. For solids, one speaks typically of the surface energy, surface free energy, or the excess surface free energy, whereas for liquids one uses surface tension. In all cases, all of these terms refer to the extra energy per unit area required to form a surface of a condensed phase (solid or liquid) that is exposed to a rarefied phase (gas or vacuum). This will have units of free energy per unit area, or equivalently, a force per unit length. The nature of this actual force exerted normal to a unit line will be discussed below. For the purpose of this chapter, we shall abide by common practice and refer to the surface free energy per unit area as the surface energy in the case of solids, and the surface tension in the case of liquids. Note that, if both phases are condensed (that is, solid–solid, solid–liquid, liquid–liquid) then we refer to the interface free energy. Why is it a free energy? Because that is the thermodynamic energy that can be converted to work at a finite temperature. Thus, the surface energy involves contributions arising from internal energy and from entropy. Why does a surface (or an interface) have an additional or excess energy compared to an infinite solid? Because it is a defect. The atoms comprising the solid would like to be bonded to other atoms, which lowers their free energy. The bonds that are broken by forming a surface store energy, and this corresponds to the excess free energy of the surface.

4.3   SURFACE TENSION: LIQUIDS

The reversible thermodynamic work per unit required to create the free surface is what we define as the surface tension. It is often denoted by the symbol γ, and is given by

$$\gamma = \frac{dW}{dA}$$

where W is the work and A is the surface area. The surface tension will generally be higher for stronger bonding: typical values are 10–20 mJ/m^2 for fluorocarbons, 20–40 mJ/m^2 for hydrocarbons, 73 mJ/m^2 for water, and 485 mJ/m^2 for mercury. As the temperature rises, the surface tension of most liquids goes down, reaching zero at the point of evaporation.

The term *tension* implies that a force exists, and indeed this is a real tensile force confined to the surface region. For a liquid, the force is isotropic and acts within the plane of the surface. This is what causes liquid droplets to adopt spherical shapes, because that is the minimum surface area configuration for a given volume. The tensile force arises because, compared with atoms in the bulk, the atoms at the surface are lacking bonds with neighbors outside of the volume of the liquid.

Because of the tensile force, one can think of the surface as acting like a stretched elastic membrane, an effect that is readily observed, for example, when light objects that repel water rest on the water surface, enabling the water strider insect (Figure 4.4) to skim, instead of having to swim!

The surface tension leads to a real force that can indeed act on other objects and can be measured. Figure 4.5 illustrates a thin liquid film suspended within a C-shaped frame and a movable slider. The slider has a length L. The liquid has both an upper and a lower surface in contact with the slider, as shown in the side view free body diagram. Each surface exerts a force per unit length γ on the wire, so the total force is $2\gamma L$ acting to the left. A force $F_{applied}$ equal in magnitude and opposite

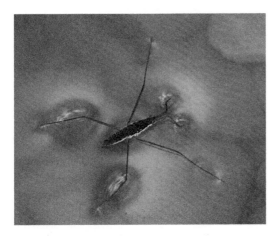

FIGURE 4.4   A water strider. The membrane-like nature of water due to its surface tension is clearly seen. The behavior is assisted by the hydrophobic nature of the insect's tentacles. Public image from http://en.wikivisual.com/index.php/Image:Waterstrider.jpg.

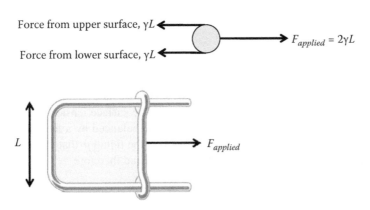

FIGURE 4.5   The surface tension of a liquid can be measured using a C-shaped wire frame, with a slider attached to it. The inset shows a free body diagram of a cross section of the slider illustrating the forces acting on the wire: the applied force F, and the two forces from the two surfaces of the liquid. Adapted from (permission pending): John D. Cutnell and Kenneth W. Johnson, *Essentials of Physics* (New York: Wiley Publishing, 2006).

in direction must be applied to counteract this force. Here we see illustrated a critical consequence of surface tension when a liquid–vapor interface makes contact with another object: the surface produces a force per unit length, where the force is directed in the plane of the surface normal to the line of contact with the object, and the length is measured along the line of contact. This force enables the surface tension of a liquid to be measured experimentally—for example, by the Whilhelmy plate method [6].

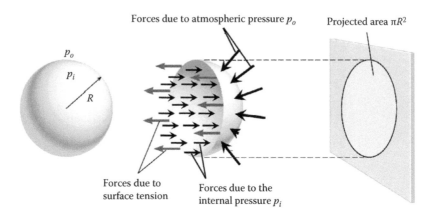

Forces due to atmospheric pressure p_o Projected area πR^2

p_o

p_i

R

Forces due to
surface tension

Forces due to the
internal pressure p_i

FIGURE 4.6 A liquid droplet is shown on the left. On the right is a free body diagram cut through the droplet. The surface tension acts around the exposed rim of the surface (light gray arrows). The internal pressure of the droplet acts to the right through the exposed cross section. The external pressure acts on the outer surface. The net force from this external pressure is proportional to p_o multiplied by the projected area πR^2. Adapted from (permission pending): John D. Cutnell and Kenneth W. Johnson, *Essentials of Physics* (New York: Wiley Publishing, 2006).

4.4 LAPLACE PRESSURE

Consider a spherical liquid droplet of radius R (Figure 4.6). We slice the droplet in half and draw a free body diagram of one half. The surface tension produces a net force to the left of magnitude $2\pi R\gamma$. This must be balanced by a force to the right. The force must come from an internal pressure in the liquid p_i that produces a force $\pi R^2 p_i$. At equilibrium, these forces must be equal, and therefore

$$p_i = \frac{2\gamma}{R}$$

If there is an external pressure p_o, then elementary analysis shows that the pressure difference between the outside and inside $\Delta p = p_i\text{-}p_0$ must correspond to this amount. That is,

$$\Delta p = \frac{2\gamma}{R}$$

This pressure is called the Laplace pressure, or capillary pressure. In this case, the pressure is positive, representing a compressive pressure inside the liquid. The pressure can change sign if the curvature radius R of the liquid–vapor interface changes from positive curvature (i.e., convex as viewed from outside the liquid) to negative (i.e., concave as viewed from outside the liquid), like one might see at the rim of a glass or at the top of a capillary tube when the liquid wicks upward along the surface

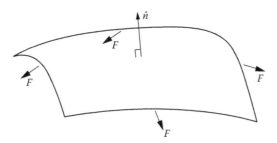

FIGURE 4.7 Free body diagram of a surface with two distinct arbitrary curvature radii, which are referred in the text as R_x and R_y. Surface tension forces are indicated. Force equilibrium requires an internal pressure given by Equation (4.1).

of the container. The reduced pressure with respect to the outside vapor explains the commonly observed effect of the rise of a liquid in a small capillary tube inserted into a liquid reservoir. More generally, the capillary pressure is given by:

$$\Delta P = \gamma \left(\frac{1}{R_1} + \frac{1}{R_2} \right) = 2\gamma\kappa = \gamma \, \nabla \cdot \hat{n} \qquad (4.1)$$

where R_1 and R_2 are the principal radii of curvature of the liquid–vapor interface at the point of interest, κ is the corresponding mean curvature, and $\hat{n}$ is the normal vector to the surface (pointing from the liquid outward into the vapor) (Figure 4.7). This is called the Young-Laplace equation [7,8].

As we will see later, the sign of the curvature will be affected by the contact angle the liquid makes with a solid, which is determined by the tendency of the liquid to wet the surface. For a pair of closely-spaced parallel surfaces, like a cantilever in a MEMS device that hovers over a surface (Figure 4.8), liquids that tend to wet the surfaces will form a capillary with negative curvature surfaces. This creates a negative capillary pressure and thus a tensile force that pulls the cantilever downward toward the surface. This can lead to the cantilever remaining stuck on the surface, even after the liquid is removed due to subsequent solid–solid adhesion. This can be a common occurrence when a MEMS device is removed from a liquid processing step during fabrication, such as when the sacrificial layer between the cantilever is chemically etched to remove it. For a MEMS device, this essentially renders it non-functional. On the contrary, if the surfaces tend to repel the liquid, the capillary surfaces will have positive curvature, producing a force that pushes the surfaces apart, preventing contact from occurring. This effect is now used routinely in the fabrication of MEMS devices, where the surfaces are made to repel water and other liquids through organic coatings [9].

As another example, consider a round asperity in contact with a flat surface (Figure 4.9). Consider a liquid that tends to wet the surfaces that has condensed at the contact. At each point on the liquid–vapor interface, the two principal radii of curvature have opposite sign. Figure 4.9 shows an approximate representation of these radii; an exact treatment is given elsewhere [10] (the principal curvature radii

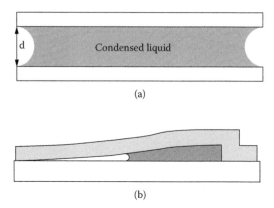

(a)

(b)

FIGURE 4.8 Schematic of (a) capillary condensation between two parallel plates (side view) and (b) a microstructure after drying, where the process has led to the compliant structure adhering to the substrate. Permission to reproduce pending from R. Maboudian and R.T. Howe, "Critical Review: Adhesion in Surface Micromechanical Structures." *J. Vac. Sci. Technol.* 15 (1997): 1–20, Figure 2.

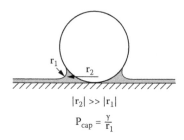

$$|r_2| \gg |r_1|$$

$$P_{cap} = \frac{\gamma}{r_1}$$

FIGURE 4.9 A capillary condensed at a sphere–plane contact. Permission to reproduce pending from C.M. Mate, *Tribology on the Small Scale: A Bottom Up Approach to Friction, Lubrication, and Wear* (New York: Oxford University Press, 2008), Figure 5-6(c).

vary at all points on the meniscus surface, and the values illustrated in Figure 4.9 are approximations of their average values). Because of the geometry, the negative curvature radius r_1 has a much smaller magnitude and therefore dominates in determining the Laplace pressure. Thus, the Laplace pressure is negative and the effect of the capillary is to pull the asperity and the flat surface together, creating adhesion that will result in a contact area under compressive stress, and the adhesion must be overcome if they are to be separated. In addition, the net force that the liquid surface exerts on the asperity due to the surface tension will also result in a force pulling the asperity toward the flat surface.

4.5 THE KELVIN EQUATION

For a liquid–vapor interface with curvature under isothermal conditions, equilibrium thermodynamics leads to the Kelvin equation:

$$R_M T \ln\left(\frac{p}{p_0}\right) = \gamma V_M \left(\frac{1}{r_1} + \frac{1}{r_2}\right) \qquad (4.2)$$

where R_M is the molar gas constant (8.314472 J•mol^{-1}•K^{-1}), T is the temperature, p is the pressure of the vapor, p_0 is the bulk saturation pressure of the liquid, and V is the molar volume of the liquid (18 cm^3 for water). The ratio p/p_0 corresponds to the relative vapor pressure of the liquid, which for water is simply the relative humidity (RH). We can define the Kelvin radius r_k as:

$$r_K^{-1} = \frac{1}{2}\left(\frac{1}{r_1} + \frac{1}{r_2}\right) = \frac{R_M T}{2\gamma V_M} \ln\left(\frac{p}{p_0}\right) \qquad (4.3)$$

The quantity p/p_0 is less than one (except in supersaturation conditions), and so the Kelvin radius will be negative. The fundamental concept here is that curved surfaces have higher energy than flat surfaces for both vapors and liquids. The saturation vapor pressure over a convex (as viewed from the vapor) liquid surface is lower than the bulk saturation vapor pressure. This means that condensation from vapor can occur below saturation conditions if it is able to form a surface with negative curvature due to geometry. It also means that smaller droplets will evaporate faster than larger ones.

The Kelvin radius varies logarithmically with the partial pressure of the species. A graph of r_K versus p/p_0 for water at 20°C (where $\gamma V/R_M T = 0.54$ nm) is shown in Figure 4.10. Up to $p/p_0 = 0.75$ and lower $|r_K| < 2$ nm, showing that for water, capillary condensation is a nanoscale phenomenon that will have significant consequences for atomic force microscope (AFM) and other nanoscale applications involving surfaces in close proximity.

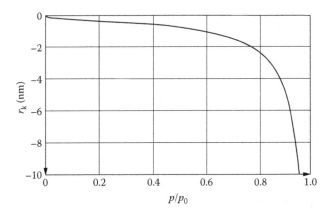

FIGURE 4.10 The Kelvin radius of water at 20°C, plotted as a function of relative humidity. Permission to reproduce pending from R.W. Carpick, J.D. Batteas, and M.P. de Boer, in *Handbook of Nanotechnology*, 2nd ed., edited by B. Bhushan, 1–2 (New York; Springer-Verlag, 2006), Figure 32.5.

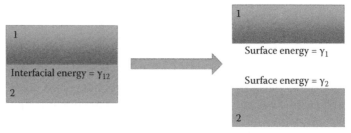

Work of adhesion $W = \gamma_1 + \gamma_2 - \gamma_{12}$

FIGURE 4.11 Schematic illustrating the change in energy per unit area upon separating an interface to form two free surfaces.

4.6 INTERFACIAL ENERGY AND WORK OF ADHESION

Consider two condensed phases 1 and 2 of respective surface energies γ_1 and γ_2. If they form a stable interface, then they will possess a distinct interfacial energy. Therefore, as illustrated in Figure 4.11, the free energy of the system changes if the two phases are separated by an amount:

$$W_{12} = \gamma_1 + \gamma_2 - \gamma_{12} \tag{4.4}$$

In most cases, W is positive (the sum of the free energies of the two surfaces are higher than the interfacial energy) and so work must be done to separate them. This quantity is called the work of adhesion, or sometimes the Dupré energy of adhesion. Some texts call this quantity γ; we will use W to distinguish it clearly from surface or interfacial free energies. It represents the reversible work done to separate the surfaces from contact to infinity, and therefore depends on the interfacial forces that act between the two materials as well as their individual surface energies.

If one is separating a single material—that is, material 1 and 2 are identical—then the interface has no additional free energy and the equation reduces to $W_{11} = 2\gamma_1$. Here, W is called the work of cohesion. Note that here γ_1 refers to the surface energy in the environment in which the surfaces are separated.

The interfacial energy can depend on the bonding and hence the precise atomic arrangement at the interface, and so variations in the atomic configuration can lead to changes in the work of adhesion. Recent atomistic modeling work has shown that significant differences in the work of adhesion can result when one changes the relative alignment, or the degree of amorphization, of two otherwise identical materials [11].

4.7 SURFACE ENERGY OF SOLIDS

4.7.1 SURFACES AS DEFECTS

A solid surface can be regarded as a type of material defect, since the bonding that is expected in a uniformly-bonded solid is interrupted. Like vacancies or interstitials (0-D defects) and dislocations (1-D defects), surfaces (2-D defects) have an additional free energy associated with them due to the interruption of the equilibrium bonding configuration; this is also essentially true for liquids. So while both solid and liquid surfaces have an energy associated with them, solid surfaces are very different from liquid surfaces, for multiple reasons. First, the strong bonding present in solids inhibits the kinetics needed for atoms to find their lowest energy configurations. Therefore, solid surfaces are rarely at their equilibrated state, and so many of the structural configurations found are often metastable. Second, liquids cannot sustain strain except for uniform (hydrostatic) strain. Atoms at a liquid surface experiencing strain will rearrange to maintain the surface energy at its original value, whereas such atomic rearrangements at a solid surface have energy barriers to overcome that can be much higher than available thermal energies, so they cannot easily rearrange at a solid surface. Elastic and plastic strain can therefore both lead to a change in the surface energy with the surface area A; that is, $d\gamma/dA \neq 0$. The history of creating the surface therefore matters.

Furthermore, the surface created by simply separating a solid along a plane, even a high symmetry plane in a crystal, is not necessarily the lowest energy surface. The point to realize is that the atoms at a surface are generally under-coordinated and therefore in a higher energy state than when they are bonded inside the uninterrupted solid. The processes occurring at a freshly prepared surface, such as one produced by brittle cleavage, are all driven by thermodynamics. Atoms will tend to rearrange to make up for this lack of coordination. For example, relaxation and reconstruction of the surface often occur, as well as chemical reactions with species in the environment. Some of these processes require sufficient thermal energy, while others may be spontaneous.

4.7.2 Relaxation

Relaxation refers to the motion of atoms inward or outward from a surface plane. This means there is a change in the average interlayer spacing, and the changes in bond lengths may be different for different atomic species. Relaxation usually affects the first 1–3 layers of a crystal, with the strongest effect at the topmost layer. It is more frequently observed with metallic and ionic materials, with the effect being much weaker in ionic materials.

In a metallic material, relaxation can be understood as resulting from two different effects. The lack of ionic cores above the solid surface allows the nearly free elections of the metal to "spill out" away from the surface due to their kinetic energy. The ionic cores of the topmost surface layer therefore find it energetically favorable to partially follow suit, relaxing away from their neighbors below. Alternately, the lack of electron density at the surface can lead to atoms relaxing inward, reducing bond lengths with the atoms in the layer below. In ionic solids, the change in electrostatic field due to the absence of atoms above them tend to cause surface atoms to relax in different directions (i.e., toward or away from) and by different amounts depending on their ionicity.

4.7.3 RECONSTRUCTION

Reconstruction refers to the rebonding of surface and possibly sub-surface atoms. The re-arrangements are both parallel and perpendicular to the surface plane, unlike relaxation that mostly involves only motions perpendicular to the surface. Because of the motions parallel to the surface plane, the surface lattice is not in registry with the bulk, unlike with relaxation. The new symmetry of the surface lattice leads to the use of new surface lattice vectors to describe the surface unit cell symmetry [12,13].

Reconstruction is more frequently observed with covalent materials, but is also sometimes observed with metals. The example of the diamond (111) surface is shown in Figure 4.12(a). The reconstruction effect is driven thermodynamically: cleaving or somehow creating a surface produces unsaturated bonds. By rearranging themselves and forming other bonds with locations and directions different from the bulk arrangement, the surface energy is lowered. Surface reconstructions can be altered or completely removed by terminating the surface with adsorbates, as shown, for example, for the diamond (111) surface in Figure 4.12(b).

4.7.4 RULES OF THUMB FOR SURFACE ENERGY

One can use simple rules of thumb to estimate how large the surface energy will be if a certain plane within the bulk of a crystal is selected for producing a surface. In particular, higher surface energy will result from: (1) a higher bond density (number of bonds per unit area) with a with a bond direction having a component normal to that surface; and (2) a higher bond energy for such bonds. These effects are counteracted by surface relaxation, reconstruction, and adsorption.

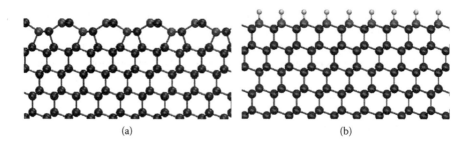

(a)   (b)

FIGURE 4.12 Surface reconstruction of the diamond (111) surface. (a) After cleaving diamond to expose a (111) plane, the surface reconstructs to reduce the surface energy. The resulting structure involves rehybridized bonds forming with 2×1 symmetry. (b) If the surface is exposed to atomic hydrogen, then under equilibrium conditions the reconstruction is lifted and a 1×1 surface terminated with hydrogen atoms is formed. Adapted from D.W. Brenner and J.A. Harrison, "Atomic simulations of diamond films." *Ceramic Bulletin* 71 (1992), 1821–1828.

4.8 CONTACT ANGLE

One important macroscopic manifestation of surface energies that is readily visible by eye is the contact angle that a liquid droplet forms when in contact with a solid surface. The contact angle is defined as the angle through the liquid, measured between the tangent line surface along the surface–liquid interface, and the tangent line along the liquid–vapor interface. Both tangent lines are drawn at the point on the cross-section through the system (normal to the solid surface) where the solid, liquid, and vapor all meet (Figure 4.13).

Young derived a relation between the contact angle and the interfacial energies involved in the system:

$$\gamma_{LV} \cos \theta = \gamma_{SV} - \gamma_{SL} \qquad (4.5)$$

This is called Young's equation, and it can be simply derived by considering the fact that the vectors shown in Figure 4.13 represent vectors of force (actually force per unit length, intersecting the plane of viewing) acting at the liquid-solid-vapor contact line. Applying force equilibrium immediately leads to Equation (4.5). A contact angle of 0° corresponds to the liquid fully spreading over the surface (full wetting), while a contact angle of 180° represents complete non-wetting, where a full droplet would be present on the surface. In the case of water droplets, surfaces that lead to droplets with contact angles below 90° are called hydrophilic, above 90°, hydrophobic. The driving force to full wetting is seen for the case of 0° contact angle. In this case, Equation (4.5) reduces to $\gamma_{LV} = \gamma_{SV} - \gamma_{SL}$, thus $\gamma_{LV} < \gamma_{SV}$; a surface fully covered with the liquid will have a lower energy than one where the solid is exposed. Thermodynamics drives the liquid to spread over the entire surface. Defects on surfaces can cause contact angle hysteresis, where the thermodynamic spreading is inhibited by pinning that occurs between the liquid and high-energy sites on the surface. This can lead to different values for advancing and receding contact angles; this can be a useful probe of surface defects. Surfaces that are almost completely non-wetting can be made by coating the surface with low-energy groups that are also unfavorable to interactions with the liquid. Endowing a flat surface that is already non-wetting with topographic roughness also can lead to an apparently larger macroscopic contact angle. For

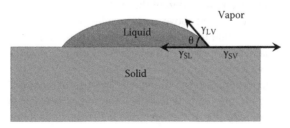

FIGURE 4.13 A liquid droplet on a solid surface. Drawing the surface tensions as vectors along the respective interfaces at a point on the solid-liquid-vapor contact line enables Young's equation (Equation [4.5]) to be derived.

(a) (b)

FIGURE 4.14 (a) Photograph of a water droplet on a superhydrophobic lotus leaf surface: the droplet touches at only a few points and forms a ball, which easily rolls off the surface with the slightest tilt. (b) Scanning electron micrograph of the surface of a lotus leaf. The surface exhibits roughness on multiple lengths scales. The combination of the hierarchical surface roughness and hydrophobic (waxy) surface chemistry leads to a macroscopic water contact angle near 180°. Permission to reproduce pending from B.N.J. Persson, O. Albohr, U. Tartaglino, A.I. Volokitin, and E. Tosatti, "On the Nature of Surface Roughness with Application to Contact Mechanics, Sealing, Rubber Friction and Adhesion." *J. Phys., Condens. Matter.* 17 (2005): 1–62, Figure 2, 3.

the case of water, surfaces where the contact angle approaches 180° are called superhydrophobic. The combination of surface chemistry and surface topography is seen in the famous case of the "lotus leaf effect" (Figure 4.14) [14]. Contact angles can be directly measured using side-view optical microscopy, with such an instrument referred to as a contact angle goniometer. Measuring the contact angle between a surface and a series of liquids can be used to determine the surface energy of the solid.

4.9 THE RELATION BETWEEN SURFACE FORCES AND SURFACE ENERGIES

Determining the forces of interaction and the resulting stresses that occur is important for nanotribological studies. An important starting point to consider is the force required to separate two surfaces that have some mutual attraction. The interatomic forces occurring and the configuration of the surfaces will lead to surface and interfacial energies, and, as discussed above, a given value for the work of adhesion. How this is manifested in an actual force required to separate the surfaces depends on the shape and the mechanical properties of the bodies in contact, as well as on the spatial range of the attractive forces.

The Derjaugin approximation is a simple but very useful relation that can be used for this. The Derjaguin approximation relates the forces between two curved surfaces F to the work of adhesion W between two flat surfaces as follows:

$$F_{sphere-on-sphere} = 2\pi\left(\frac{R_1 R_2}{R_1 + R_2}\right)W_{flat-on-flat} \qquad (4.6)$$

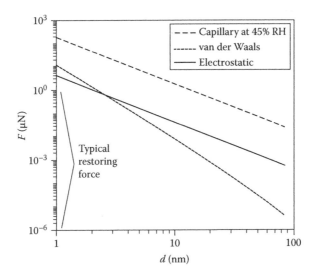

FIGURE 4.15   Comparison of attractive forces due to capillary condensation (at 45% relative humidity), van der Waals forces, and an electrostatic force for 1 V potential difference, for two 1-μm² perfectly smooth Si surfaces as a function of the separation d between them. The forces increase significantly as the gap decreases, and at gaps below ~5 nm these forces exceed typical restoring forces that can be exerted in MEMS devices. Permission to reproduce pending from R. Maboudian and R.T. Howe, "Critical Review: Adhesion in Surface Micromechanical Structures." J. *Vac. Sci. Technol.* 15 (1997): 1–20, Figure 4.

The point is that the force between curved surfaces is straightforward to measure, such as in an AFM or adhesion experiment, and is also relevant for understanding adhesion between asperities. On the other hand, the interaction forces and the work of adhesion are easier to calculate. Equation (4.6) assumes that the surfaces are at their equilibrium separation, although it can be expressed for any separation between the surfaces. It is also assumed that the range of the interaction forces are much smaller than the radii of curvature R_1, R_2. For the special case of a sphere of radius R in proximity to a flat surface, we have $R_1 = R$, $R_2 = \infty$ and therefore:

$$F_{flat-on-flat} = 2\pi W_{flat-on-flat} \tag{4.7}$$

This simple relation enables the magnitude of adhesive forces between asperities to be immediately estimated. For example, consider a diamond-like carbon (DLC) tip of radius 10 nm brought into contact with a flat DLC sample of identical composition. Assume the surface energy of this particular DLC is measured to be $\gamma = 0.035$ J/m². Assuming the tip and sample surfaces are clean and can thus be brought together to form an ideal, low- energy interface, the work of cohesion will then be $W = 0.070$ J/m². The force holding the tip and sample together will then be $F_{tip-sample} = 2\pi RW = 2\pi(10)$ $(0.070) = 4.40$ nN. Two metal surfaces that like to form covalent bonds may have a work of adhesion in the vicinity of 1–3 J/m², leading to a tip–sample force (also for a 10 nm radius tip) of approximately $63 - 188$ nN.

4.10 ADHESION IN DRY ENVIRONMENTS

The Derjaugin approximation ignores the possibility that the contacting materials may deform elastically in response to the attractive forces. This in turn can affect the force required to separate the surfaces. The problem of adhesion between elastic spheres has been thoroughly modeled and the essential results are summarized here.

We start by assuming that the contact is between a spherical tip and a flat surface. Both surfaces are perfectly smooth and frictionless, and possess a work of adhesion W. The contact radius formed due to the action of the adhesion forces is assumed to be small compared to the radius of the tip. Thus, the sphere can be approximated as having a paraboloidal shape. Finally, we assume the materials are homogeneous and isotropic with linear elastic behavior. It is important to realize that all of these assumptions may not be satisfied in an actual AFM experiment, and these assumptions should be carefully considered and tested if possible.

Within the context of these assumptions, the contact behavior falls somewhere between the limiting cases of the Johnson-Kendall-Roberts (JKR) model [15] (for large tips and compliant materials with strong, short-range adhesion), to the Derjaguin-Müller-Toporov (DMT) model [16] (for small tips and stiff materials with weak, long-range adhesion). W is related to the force F_{PO} required to pull the tip out of contact with the surface by

$$W = \frac{-F_{PO}}{\chi \pi R}$$

where χ ranges monotonically from 1.5 (JKR) to 2 (DMT). Tabor's parameter μ_T can be used to select the value between these two limits that applies [17,18]. This parameter is given by:

$$\mu_T = \left(\frac{16RW^2}{9K^2 z_0^3} \right)^{1/3} = \left(\frac{16F_{PO}^2}{9\chi^2 \pi^2 RK^2 z_0^3} \right)^{1/3}$$

where R is the tip radius, and K is the contact modulus, given by

$$K = \frac{4}{3} \cdot \left(\frac{1-v_1^2}{E_1} + \frac{1-v_2^2}{E_2} \right)^{-1}$$

E_1, E_2 are the Young's moduli, and v_1, v_2 the Poisson's ratios of the tip and sample, respectively. The parameter z_0 is the equilibrium separation of the surfaces in contact, and is assumed to represent the length scale of the interfacial forces. $\mu_T > 5$ ($\mu_T < 0.1$) implies the JKR (DMT) limit. Unfortunately, neither z_0 nor χ are not known a priori. However, an iterative method described previously [19] allows one to perform reasonable estimates. In addition to providing a relationship between adhesion forces and the geometry and material properties, these models also provide the contact area, contact stresses, contact stiffness, and deformations. The fact that the relationship between the work of adhesion and the pull-off force does not directly depend on the elastic properties (it only depends on them weakly through χ) is a special result that occurs for a paraboloidal tip. If the tip has a different shape, then the equations must be modified and a dependence on the elastic properties appears [20,21]. Modified models to take into account viscoelasticity [22–28], plasticity [29–30], anisotropy [31,32], thin film geometries [33–40], and the effect of friction and tangential forces [41,42] have been published. Surface roughness will tend to significantly decrease adhesion [43–48], although recent modeling has shed light on the ways in which interactions across small gaps between surfaces can lead to significant adhesive interactions [49]. Recently, atomistic simulations have indicated that the continuum models can break down at the nanometer scale [50,51], but in some cases they provide good approximations. This remains an active area of inquiry with substantial consequences for nanotribology measurements.

4.11 ADHESION IN WET ENVIRONMENTS

We now consider the effect of the presence of liquids on adhesion forces. The study of how liquids adsorbed onto surfaces affect interfacial behavior, including adhesion and friction, is a large and diverse area of research. Such behavior has importance in broad technological areas, including textiles, paints, lubricants, geology, environmental chemistry, and all aspects of biology.

In considering how liquids affect adhesion, two key situations emerge: (1) full immersion in a liquid; and (2) adhesion in the presence of condensable vapors, which can form a liquid meniscus. The full immersion of a pair of solids 1 and 2 into a liquid is simple to treat. The separation of the surfaces still involves the destruction of the interface 12, but now instead of creating free surfaces 1 and 2, two liquid–solid interfaces 1l and 2l are created. Hence, the work of adhesion becomes:

$$W_{12} = \gamma_{1l} + \gamma_{2l} - \gamma_{12} \tag{4.8}$$

This can lead, for example, to a substantial reduction in the pull-off force if the liquid–solid surface tensions are small (i.e., the liquid likes to wet the solids, making the separation of the solids less thermodynamically unfavorable).

In a crack, at a sharp corner, or between two closely-spaced surfaces, liquids can condense to form a meniscus. This phenomenon occurs because the saturation vapor pressure over a curved surface is different from that for a flat surface. In particular, if the liquid–vapor interface has convex curvature (as viewed from the vapor looking at the liquid), then the saturation vapor pressure is lower. Thus, a geometry that can accommodate such a meniscus (such as a narrow gap between surfaces having low contact angles with the liquid) can form a thermodynamically stable liquid meniscus.

This phenomenon is readily detected in atomic force microscopy, where a meniscus between AFM tips and samples can readily form [52,53]. In fact, any force measurement carried out in ambient laboratory conditions requires consideration of the effect of such a capillary neck forming between the tip and sample. Studying such necks can provide insight into the properties of the liquid, particularly under confined nanoscale conditions. For background, Israelachvili [7] and de Gennes [54] discuss the terminology, physics, and chemistry of liquid films and their wetting properties in detail. Adhesion in the context of AFM is discussed by Carpick, Batteas and de Boer [10].

The typical starting point involves considering the tip as an ideal sphere of radius R and applying the classical theory of capillary condensation between a sphere and a plane, as derived from the thermodynamics of capillary formation with further simplifying geometrical assumptions. If the radius of the sphere is large with respect to the size of the capillary, one can use the circle approximation, whereby radii of curvature of the meniscus are taken to be constant. The geometry of the capillary meniscus for the general case, not using the circle approximation illustrated in Figure 4.9, is now considered in Figure 4.16, which is adapted from de Boer and de Boer [55]. Instead of the assuming that the radii r_1 and r_2 are constant, we consider the principal radii of curvature at any point on the meniscus surface. These are referred to as the azimuthal radius r_a, and the meridional radius, r_m. These vary locally, but still satisfy the Kelvin equation (Equation 4.2). The quantity D represents the separation between the tip and sample, and the angle ψ is referred to as the filling angle.

The resulting tip–sample forces come from two sources: (1) the Laplace pressure, and (2) the liquid–vapor surface tension. For both sources, the geometry of the meniscus must be determined. In the general case, this is done through numerical integration. If one uses the circle approximation and if the two contact angles are known, then the only unknowns in the problem will be r_1 and r_2. The circle approximation creates one constraint for these two quantities, and the other, assuming equilibrium conditions, is provided by the Kelvin equation (Equation 4.2).

One problem with the circle approximation is that it assumes the circle radius r_1 to be constant everywhere. The Kelvin equation then forces r_2 to be constant everywhere, but the geometry in Figure 4.16 shows clearly that the interior width of the meniscus, which corresponds to the radius r_2, is very different at different points on the meniscus. In other words, the circle approximation creates a situation that is not isobaric.

The Laplace pressure is given by Equation (4.1). If $\Delta p < 0$, then the resulting force between the tip and sample is attractive. This is the case for Figure 4.16. In the circle approximation, one would have $r_1 < 0$, $r_2 > 0$, and $|r_1| = |r_2|$. The maximum attractive force F_c between the tip and sample due just to the Laplace pressure is readily shown to be:

$$F_C = -4\pi\gamma R\cos\theta \qquad (4.9)$$

if we assume that $\theta_1 = \theta_2$ [7,8,56,57]. This predicts that the adhesion force is independent of the partial pressure of the vapor, or in the case of water, the relative humidity. Thus, a finite adhesion force is predicted at 0% RH, when in fact there should be no capillary formation. Despite this problem, Equation (4.9) is commonly used to interpret adhesion AFM experiments. There are several assumptions built in to Equation (4.9), some of which have already been mentioned. These assumptions are:

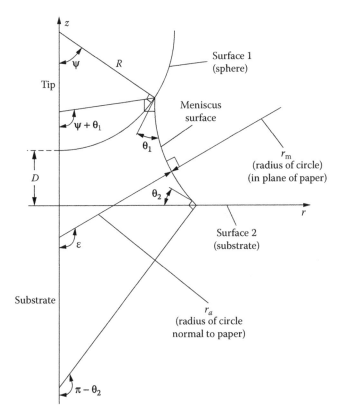

FIGURE 4.16 The AFM tip considered as a sphere of radius R at a distance D from the sample. The liquid film in between forms a meniscus with principal curvature radii r_a (the azimuthal radius) and r_m (the meridional radius) indicated for one point on the meniscus surface. The contact angles formed with the tip and substrate are indicated along with other key angles used in the analysis of the problem. Adapted from (permission pending) M.P. de Boer and P.C.T. de Boer, "Thermodynamics of Capillary Adhesion between Rough Surfaces." *J. Colloid Interface Sci.* 311 (2007): 171–185, Figure 8.

(a) $|r_1| \ll R$, which is equivalent to saying ϕ is small

(b) $|r_1| \ll |r_2|$.

(c) $\theta_1 = \theta_2$.

(d) $D \ll |r_1|, |r_2|$.

(e) The tip is shaped like a perfect sphere.

(f) The solid–solid adhesion force is negligible with respect to the meniscus force.

(g) The meniscus cross sections are perfect circular arcs (the circle approximation).

(h) The force from the Laplace pressure dominates the force due to the resolved surface tension of the meniscus.

(i) The liquid–vapor surface tension γ is independent of the meniscus size.

(j) The meniscus volume remains constant as the tip is retracted.

(k) The maximum force of attraction is equal to the pull-off force.
(l) The tip and sample are rigid.

Importantly, several of these may not be correct for AFM experiments or other nanoscale asperities. These are discussed in more detail elsewhere [10]. Corrections to account for assumptions (a)–(h) were worked out by Orr and Scriven [57].

Assumptions (i) and (j) are not assumptions of Equation (4.9) itself, but rather they are assumptions that are often used when applying Equation (4.9) to AFM measurements. Equation (4.9) simply gives the maximum force of attraction between the tip and sample. As discussed previously, an AFM does not measure this quantity. Rather, it measures the force at which instability occurs. If a capillary has formed between the tip and sample, then the force as a function of distance can be calculated. Calculating this force requires making one of two assumptions: either the volume of the capillary is conserved (because the rate of displacement is large with respect to the adsorption or desorption kinetics of the liquid) or the Kelvin radius is conserved (i.e., the rate of displacement is slow with respect to the adsorption or desorption kinetics of the liquid, and so the capillary remains in equilibrium). The constant volume assumption (j) has been used in almost all published work. Israelachvili, however, pointed out the difference between these two approaches in his book [7], and left the solution of the problem as an exercise to the reader. The force is the same at $D = 0$, but the reduction in force with displacement is more rapid and linear for the constant Kelvin radius case. With the assumptions listed previously for the constant volume case

$$F(D) = 4\pi R \gamma_l \cos \theta \left(1 - \frac{D}{\sqrt{4r_k^2 \cos^2 \theta + D^2}} \right) \qquad (4.10)$$

while for the constant Kelvin radius case,

$$F(D) = 4\pi R \gamma_l \cos \theta \left(1 - \frac{D}{2|r_K| \cos \theta} \right) \qquad (4.11)$$

As with the problem of scale-dependent surface tension mentioned previously, the kinetics of capillary formation and dissolution is a relatively unexplored problem and is therefore worthy of further investigation. A recent study of the humidity dependence of friction as a function of sliding speed is an example where this issue is raised [58].

Once an assumption about how the meniscus changes with displacement has been made, one still needs to consider the nature of the instability in order to relate the AFM pull-off measurement to the capillary's properties. As stated previously and shown in Figure 4.2, a low lever stiffness k or a strongly varying adhesive force will lead to a pull-off force that is nearly equal to the adhesive force, and so assumption (k) would be valid. However, if k is sufficiently large, or the capillary stiffness is sufficiently weak, this assumption will fail. As we shall see in the following text, experimental efforts to investigate this point have yet to be carried out.

Finally, assumption (l), if violated, requires a substantially more complex analysis. The question has been addressed independently by Maugis and Gauthiermanuel [59] and Fogden and White [60]. Both papers provide a non-dimensional parameter that allows one to determine the severity of the effect. In the limit of small tips, stiff materials, large (in magnitude) Kelvin radii, and low surface tensions, the effect of elastic deformation is negligible. However, for relatively compliant materials, large tips, and small Kelvin radii, the meniscus can appreciably deform the contact in the immediate vicinity of the meniscus. This can substantially alter the mechanics of adhesion as well as significantly affecting the stresses. The dependence on the Kelvin radius is particularly critical. This effect may be of particular concern with soft materials like polymers or biological specimens. According to Maugis, the problem becomes analogous to the adhesive contact problem for solids, discussed by Johnson, Kendall, and Roberts [15] and further studied in many papers since [17,61,62].

Recently, calculations were presented that take into account the factors above [10,55,63]. Several important results emerge from the analysis. First, for a particular case of water between two fully hydrophilic surfaces where the sphere radius is 20 nm (comparable to many AFM tips), the maximum attractive force is smaller, but within 20% of the prediction of Equation (4.9) up to $p/p_0 = 0.9$, after which it decreases more substantially [10]. The decrease is due to a repulsive contribution that emerges from the reduction of the Gibbs' free energy with separation, which was pointed out some years ago [53,64].

Second, and more generally, the force versus separation profile is non-monotonic, exhibiting an unstable low-force branch. Moreover, the point of separation may correspond to a force that is very different from the maximum attractive force that occurs at the minimum separation. The point of separation is determined by the same stability criterion as any load-separation measurement: surfaces will separate when the magnitude of the negative attractive force gradient exceeds the spring constant of the cantilever. Thus, the pull-off force measured in an AFM experiment can be substantially different from the prediction of the maximum force. Measurements with cantilevers of different force constants can therefore give very different results for the pull-off force.

Third, it is important to consider the possibility of evaporation or condensation during the separation process. For a slow rupture where equilibrium conditions are attained, thermodynamics drives evaporation or condensation, and the resulting heat flow changes the work done. For a sphere on a flat, the work of adhesion is found to be half of the surface energy if equilibrium is maintained, due to evaporation [55,63]. Therefore, the speed and temperature (i.e., the kinetics) must be considered when interpreting AFM measurements. The analysis presented includes an extension of the work of Orr and Scriven for non-wetting surfaces [57]. The deviations from the approximate equations commonly used, such as Equation (4.9), are substantial.

Finally, the adhesion force can be affected if the liquid tends to wet the tip and sample materials. This has been modeled very recently, and the predictions agree well with experiments [65]. Finally, other recent work has indicated that, for the case of water, an ice-like layer can form on hydrophilic surfaces [66]. This will have significant effects on adhesion and friction. This effect, an intrinsically nanoscale interfacial phenomenon, illustrates the challenges and opportunities that arise when considering surface forces and surface energies at the nanoscale.

In summary, proper interpretation of AFM adhesion measurements requires utilization of the analyses presented in these papers.

ACKNOWLEDGMENT

This material was supported by the National Science Foundation under Grant No. 0826076.

REFERENCES

1. P.F. Vankessel, L.J. Hornbeck, R.E. Meier, and M.R. Douglass. 1998. "MEMS Based Projection Display." *Proc. IEEE*, 86: 1687.
2. R. Maboudian and R.T. Howe. 1997. "Critical Review: Adhesion in Surface Micromechanical Structures." *J. Vac. Sci. Technol.*, 15: 1–20.
3. R. Maboudian, W.R. Ashurst, and C. Carraro. 2002. "Tribological Challenges in Micromechanical Systems." *Trib. Lett.*, 12: 95–100.
4. K.J. Klabunde and R.M. Richards, (eds.) 2009. *Nanoscale Materials in Chemistry*. Hoboken, NJ: Wiley, Hoboken, N.J.
5. C.M. Mate. 2008. *Tribology on the Small Scale: A Bottom Up Approach to Friction, Lubrication, and Wear*. New York: Oxford University Press, Oxford; New York.
6. K. Holmberg, (ed.) 2002. *Handbook of Applied Surface and Colloid Chemistry*. New York: Wiley and Sons, New York.
7. J.N. Israelachvili. 1992. *Intermolecular and Surface Forces*, 2nd ed. London: Academic Press London, London.
8. T. Stifter, O. Marti, and B. Bhushan. 2000. "Theoretical Investigation of the Distance Dependence of Capillary and Van Der Waals Forces in Scanning Force Microscopy." *Phys. Rev. B*, 62: 13667–13673.
9. R. Maboudian. 1998. "Surface Processes in MEMS Technology." *Surf. Sci. Rep.*, 30: 209.
10. R.W. Carpick, J.D. Batteas, and M.P. de Boer. 2006. In *Handbook of Nanotechnology*, 2nd ed., (ed. B. Bhushan) 1–2. New York: Springer-Verlag, New York.
11. P.L. Piotrowski, R.J. Cannara, G. Gao, J.J. Urban, R.W. Carpick, and J.A. Harrison. 2010. "Atomistic Factors Governing Adhesion between Diamond, Amorphous Carbon, and Model Diamond Nanocomposite Surfaces." *J. Adhes. Sci. Technol. A*, 24: 2471–2498.
12. A. Zangwill. 1988. *Physics at Surfaces*. Cambridge, UK: Cambridge University Press, Cambridge [Cambridgeshire], New York.
13. G.A. Somorjai. 1994. *Introduction to Surface Chemistry and Catalysis*. New York: Wiley.
14. B.N.J. Persson, O. Albohr, U. Tartaglino, A.I. Volokitin, and E. Tosatti. 2005. "On the Nature of Surface Roughness with Application to Contact Mechanics, Sealing, Rubber Friction and Adhesion." *J. Phys., Condens. Matter.* 17: 1–62.
15. K.L. Johnson, K. Kendall, and A.D. Roberts. 1971. "Surface Energy and the Contact of Elastic Solids." *Proc. Roy. Soc. London A*, 324: 301–313.
16. B.V. Derjaguin, V.M. Muller, and Y.P. Toporov. 1975. "Effect of Contact Deformations on the Adhesion of Particles." *J. Colloid Interface Sci.*, 53: 314–326.

17. K. Johnson and J. Greenwood. 1997. "An Adhesion Map for the Contact of Elastic Spheres." *J. Colloid Interface Sci.*, 192: 326–333.

18. D. Tabor. 1977. "Surface Forces and Surface Interactions." *J. Colloid Interface Sci.*, 58: 2.

19. D.S. Grierson, E.E. Flater, and R.W. Carpick. 2005. "Accounting for the JKR-DMT Transition in Adhesion and Friction Measurements with Atomic Force Microscopy." *J. Adhes. Sci. Technol.*, 19: 291–311.

20. D. Maugis. 1995. "Extension of the JKR Theory of the Elastic Contact of Spheres to Large Contact Radii." *Langmuir*, 11: 679.

21. R.W. Carpick, N. Agraït, D.F. Ogletree, and M. Salmeron. 1996. "Measurement of Interfacial Shear (Friction) with an Ultrahigh Vacuum Atomic Force Microscope." *J. Vac. Sci. Technol. B*, 14: 1289–1295.

22. K.J. Wahl, S.V. Stepnowski, and W.N. Unertl. 1998. "Viscoelastic Effects in Nanometer-Scale Contacts under Shear." *Trib. Lett.*, 5: 103–107.

23. D. Maugis and M. Barquins. 1978. Fracture Mechanics and the Adherence of Viscoelastic Bodies." *J. Phys. D. Appl. Phys.*, 11: 1989–2023.

24. V.S. Mangipudi and M. Tirrell. 1998. "Contact-Mechanics-Based Studies of Adhesion between Polymers." *Rubber Chemistry and Technology,* 71: 407–448.

25. W.N. Unertl. 2000. "Creep Effects in Nanometer-Scale Contacts to Viscoelastic Materials: A Status Report." *Journal of Adhesion*, 74: 195–226.

26. M. Giri, D.B. Bousfield, and W.N. Unertl. 2001. "Dynamic Contacts on Viscoelastic Films: Work of Adhesion." *Langmuir*, 17: 2973–2981.

27. E. Barthel, G. Haiat, and M.C. Phan Huy. 2003. "The Adhesive Contact of Viscoelastic Spheres." *Journal of the Mechanics and Physics of Solids*, 51: 69–99.

28. K.R. Shull. 2002. "Contact Mechanics and the Adhesion of Soft Solids." *Materials Science Engineering R: Reports,* R36: 1–45.

29. B.A. Galanov, V. Domnich, and Y. Gogotsi. 2003. "Elastic-Plastic Contact Mechanics of Indentations Accounting for Phase Transformations." *Experimental Mechanics,* 43: 303–308.

30. S.D. Mesarovic and K.L. Johnson. 2000. "Adhesive Contact of Elastic-Plastic Spheres." *Journal of the Mechanics and Physics of Solids,* 48: 2009–2033.

31. D. Shi, Y. Lin, and T.C. Ovaert. 2003. "Indentation of an Orthotropic Half-Space by a Rigid Ellipsoidal Indenter." Journal of Tribology, 125: 223–231.

32. V.I. Fabrikant. 2004. "Application of the Generalized Images Method to Contact Problems for a Transversely Isotropic Elastic Layer." *Journal of Strain Analysis for Engineering Design*, 39: 55–70.

33. E.D. Reedy, Jr. 2006. "Thin-Coating Contact Mechanics with Adhesion." *J. Mater. Res.*, 21: 2660–2668.

34. E.D.J. Reedy. 2007. "Contact Mechanics for Coated Spheres That Includes the Transition from Weak to Strong Adhesion." *J. Mater. Res.*, 22: 2617–2622.

35. P.M. McGuiggan, J.S. Wallace, D.T. Smith, I. Sridhar, Z.W. Zheng, and K.L. Johnson. 2007. "Contact Mechanics of Layered Elastic Materials: Experiment and Theory." *Journal of Physics D: Applied Physics*, 40: 5984–5994.

36. Z.W. Zheng, I. Sridhar, K.L. Johnson, and W.T. Ang. 2004. "Adhesion between an AFM Probe and an Incompressible Elastic Film." *Int. J. Nanosci.*, 3: 599–608.

37. I. Sridhar, Z.W. Zheng, and K.L. Johnson. 2004. "A Detailed Analysis of Adhesion Mechanics between a Compliant Elastic Coating and a Spherical Probe." *J. Phys. D. Appl. Phys.*, 37: 2886–2895.

38. K.L. Johnson and I. Sridhar. 2001. "Adhesion between a Spherical Indenter and an Elastic Solid with a Compliant Elastic Coating." *Journal of Physics D: Applied Physics*, 34: 683–689.

39. I. Sridhar, K.L. Johnson, and N.A. Fleck. 1997. "Adhesion Mechanics of the Surface Force Apparatus." *Journal of Physics D: Applied Physics,* 30: 1710–1719.

40. I.J. Hill and W.G. Sawyer. 2010., "Energy, Adhesion, and the Elastic Foundation." *Trib. Lett.*, 37: 453–461.

41. K.S. Kim, R.M. McMeeking, and K.L. Johnson. 1998. "Adhesion, Slip, Cohesive Zones and Energy Fluxes for Elastic Spheres in Contact." *Journal of the Mechanics and Physics of Solids,* 46: 243–266.

42. K.L. Johnson. 1997. "Adhesion and Friction between a Smooth Elastic Spherical Asperity and a Plane Surface." *Proc. Roy. Soc. London A*, 453: 163–179 (1997).

43. D.L. Liu, J. Martin, and N.A. Burnham. 2007. "Optimal Roughness for Minimal Adhesion." *Appl. Phys. Lett.*, 91: 043107.

44. S. Hyun, L. Pel, J.F. Molinari, and M.O. Robbins. 2004. "Finite-Element Analysis of Contact between Elastic Self-Affine Surfaces." *Phys. Rev. E*, 70: 026117-1.

45. K.N.G. Fuller and D. Tabor. 1975. "The Effect of Surface Roughness on the Adhesion of Elastic Solids." *Proc. Roy. Soc. London A,* 345: 327–342.

46. D. Maugis. 1996. "On the Contact and Adhesion of Rough Surfaces." *J. Adhes. Sci. Technol.*, 10: 161–175.

47. T.S. Chow. 2001. "Nanoadhesion between Rough Surfaces." *Phys. Rev. Lett.*, 86: 4592–4595.

48. E.R. Beach, G.W. Tormoen, J. Drelich, and R. Han. 2002. "Pull-Off Force Measurements between Rough Surfaces by Atomic Force Microscopy." *J. Colloid Interface Sci.*, 247: 84–99.

49. F.W. Delrio, M.P. De Boer, J.A. Knapp, E.D. Reedy, P.J. Clews, and M.L. Dunn. 2005. "The Role of Van Der Waals Forces in Adhesion of Micromachined Surfaces." *Nature Mat.*, 4: 629–634.

50. B. Luan and M.O. Robbins. 2006. "Contact of Single Asperities with Varying Adhesion: Comparing Continuum Mechanics to Atomistic Simulations." *Phys. Rev. E*, 74: 26111-1.

51. B. Luan and M.O. Robbins. 2005. "The Breakdown of Continuum Models for Mechanical Contacts." *Nature*, 435: 929–932.

52. Y. Sugawara, M. Ohta, T. Konishi, S. Morita, M. Suzuki, and Y. Enomoto. 1993. "Effects of Humidity and Tip Radius on the Adhesive Force Measured with Atomic Force Microscopy." *Wear*, 168: 13–16.

53. M. Binggeli and C.M. Mate. 1994. "Influence of Capillary Condensation of Water on Nanotribology Studied by Force Microscopy." *Appl. Phys. Lett.*, 65: 415–417.

54. P.G. de Gennes. 1985. "Wetting: Statistics and Dynamics." *Reviews of Modern Physics,* 57: 827–863.

55. M.P. de Boer and P.C.T. de Boer. 2007. "Thermodynamics of Capillary Adhesion between Rough Surfaces." *J. Colloid Interface Sci.*, 311: 171–185.

56. X. Xiao and Q. Linmao. 2000. "Investigation of Humidity-Dependent Capillary Force." *Langmuir*, 16: 8153–8158.

57. F.M. Orr, L.E. Scriven, and A.P. Rivas. 1975. "Pendular Rings between Solids: Meniscus Properties and Capillary Force." *Journal of Fluid Mechanics*, 67: 723–742.

58. E. Riedo, F. Levy, and H. Brune. 2002. "Kinetics of Capillary Condensation in Nanoscopic Sliding Friction." *Phys. Rev. Lett.*, 88: 185505/1–4.

59. D. Maugis and B. Gauthiermanuel. 1994. "JKR-DMT Transition in the Presence of a Liquid Meniscus." *J. Adhes. Sci. Technol.*, 8: 1311–1322.

60. A. Fogden and L.R. White. 1990. "Contact Elasticity in the Presence of Capillary Condensation. 1. The Nonadhesive Hertz Problem." *J. Colloid Interface Sci.*, 138: 414–430.

61. J.A. Greenwood. 1997. "Adhesion of Elastic Spheres." *Proc. Roy. Soc. London A*, 453: 1277–1297.

62. K.L. Johnson. 1997. "Adhesion and Friction between a Smooth Elastic Asperity and a Plane Surface." *Proc. Roy. Soc. London A,* 453: 163–179 (1997).

63. P.C.T. De Boer and M.P. De Boer. 2008.,"Rupture Work of Pendular Bridges." *Langmuir*, 24: 160–169.

64. M. He, A.S. Blum, D.E. Aston, C. Buenviaje, R.M. Overney, and R. Luginbuhl. 2001. "Critical Phenomena of Water Bridges in Nanoasperity Contacts." *J. Chem. Phys.*, 114, 1355–1360.

65. D.B. Asay, M.P. De Boer, and S.H. Kim. 2010. "Equilibrium Vapor Adsorption and Capillary Force: Exact Laplace-Young Equation Solution and Circular Approximation Approaches." *J. Adhes. Sci. Technol.*, 24: 2363–2382.

66. D.B. Asay and S.H. Kim. 2005. "Evolution of the Adsorbed Water Layer Structure on Silicon Oxide at Room Temperature." *J. Phys. Chem. B*, 109: 16760–1676-3.

5 Nanoscale Friction
Measurement and Analysis

Rachel J. Cannara

CONTENTS

5.1 NANOTRIBOLOGY

Friction and wear are major causes of mechanical failure and dissipative energy loss. Their impact was first quantified in a study in the United Kingdom, which concluded that these losses account for 6% of the annual gross domestic product in the United States (Jost, 1966; Jost, 1976). This amounted to greater than $200 billion in 1974 and approximately $800 billion in 2010. In addition, the report predicted that tens of billions of US dollars could be saved by proper use of lubricants. In response to this need, both solid and liquid lubricants have been developed to minimize frictional energy losses, reduce equipment maintenance, and extend device lifetimes. Today, these issues are the focus of significant studies in emerging technologies involving micro- and nanoscale mechanical components and present new technical challenges for tribologists.

The principles of tribology that are relevant to the design of macroscale systems often fail at the nanoscale. This is a result of the ratio of surface to volumetric (or bulk) forces scaling inversely with an object's physical dimensions. Although the mechanical behavior of macroscopic systems is determined mainly by bulk properties and inertia, the ratio between surface and bulk forces increases as length scales decrease. One path toward demonstrating paradigm changes with reduced size is to consider the connection between frictional and gravitational forces. At the macroscale, this is represented by coefficients that relate the normal force (i.e., gravity) on a macroscopic object to friction. At the macroscale, gravitational forces tend to dominate. This is demonstrated by the example described in Figure 5.1, which compares a 10-cm cube to a 10 μm cube. As the size of the cube is reduced to 10 μm, frictional forces far exceed gravitational forces. As opposed to macroscopic coefficients of friction, the important quantity at smaller scales is the shear strength, τ, which is the *intrinsic* resistance to sliding for a given tribological pair. Expressed in units of

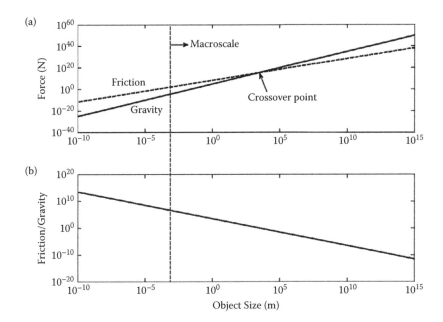

FIGURE 5.1 (a) Friction and gravitational forces are plotted as a function of object size, where the object is taken to be a steel cube with a steel-on-steel interfacial shear strength, τ, of 200 MPa and *perfectly smooth* surfaces (i.e., the calculated friction forces arise purely from adhesive load between the two objects and do not account for any surface roughness or additional load due to gravitation). Well into the macroscale, the gravitational force gives an upper limit for friction, assuming a maximum friction coefficient of 1. The crossover point indicates where friction forces begin to dominate gravity as the length scale is decreased. As an example, the force of gravity on a macroscale steel cube with 10-cm-long sides is approximately 100 N. For a 10-μm cube, the gravitational force is reduced by twelve orders of magnitude to 0.1 nN. For perfectly smooth surfaces, the friction force required to initiate sliding between two identical cubes is equal to τ times the contact area. At 10 cm, this force is 2 MN—an unreasonably high value, because it does not account for any roughness, which would reduce the actual contact area (and shift the crossover point to a smaller length scale). At 10 μm, however, the friction force is 20 mN. Despite the unreasonably high *macroscale* friction force calculated in this way and the much lower microscale friction force, the ratio of friction to gravity is still four orders of magnitude greater at the *microscale* than for the macroscale cube. This is illustrated in (b), where the ratio of friction force to gravity is shown to increase as object size decreases. In both (a) and (b), the macroscale regime is assumed to encompass all length scales > 1 mm, as indicated by the dashed vertical line.

(lateral) force per area, at its most fundamental level, τ describes the average force per interfacial atom that must be applied to initiate and maintain sliding between two perfectly smooth surfaces in contact. This paradigm shift in the relation between friction and gravity, where gravity is no longer a relevant normal force, emphasizes the importance of surface effects at reduced length scales, where surface interactions can become devastating for devices.

Although surface interactions remain important at all scales, lubrication has helped mitigate detrimental surface effects (adhesion, energy dissipation, and wear) for many macroscopic systems (Hutchings, 1992; Williams, 1994; Stachowiak and Batchelor, 2005). A lubricant acts as a buffer between surfaces in sliding contact. It produces interfaces with more favorable chemical or mechanical interactions. As a result, the lubricant–surface or lubricant–lubricant interface screens unwanted effects of a direct surface-to-surface contact. For macroscopic devices, introducing a lubricant at an interface is generally straightforward. The separations between objects in a micro- or nanoscale system, however, are much smaller and can be less amenable to stable lubrication. In addition, for liquid lubricants, the viscosity of the fluid impedes motion of micro- and nanoscale parts, and its surface tension can cause these parts to warp and adhere. Hence, while macroscale lubrication schemes often rely on the formation of a solid or liquid interface where a lubricant slides against itself, tribological principles applicable to micro- and nanoscale devices (i.e., nanotribology) must focus primarily on surface interactions at the original interface. This fundamental difference is key to understanding and controlling friction at the nanoscale.

Tribological phenomena have slowed the development of micro- and nanomechanical systems (M/NEMS), limiting their commercial application to those systems that do not include contacting sliding interfaces (Romig et al., 2003). MEMS components tend to stick together as a result of large surface forces. This "stiction" behavior poses an engineering problem both for the device itself and its process design. For example, the drying step in a liquid etch or release process forms menisci whose surface tension forces freestanding mechanical parts into irreversible contact. Researchers have improved this process by using freeze-drying, supercritical CO_2 drying, or laser heating to prevent or reverse stiction due to capillary meniscus formation (Guckel et al., 1990; Mulhern et al., 1993; Fushinobu et al., 1996; Phinney et al., 2000; Rogers and Phinney, 2002). More examples can be found elsewhere in comprehensive reviews of the tribological properties of MEMS (Maboudian and Howe, 1997; Romig et al., 2003; Maboudian and Carraro, 2003; Williams, 2006; Kim et al., 2007). Despite many important advances, failure is unavoidable for most MEMS devices (e.g., gears, locks, shutters, optical switches, etc.) that incorporate *sliding* interfaces.

The best performance to date was demonstrated by a silicon MEMS tribometer (Asay et al., 2008a, 2008b), a MEMS device designed for measuring micro- and nanoscale tribological properties. With a lifetime of at least 11 days, corresponding to 10^8 cycles, the contact traversed more than 1.5 kilometers. At that point, the test was halted, but failure was not yet observed. This extended lifetime is a result of a self-replenishing, vapor-phase lubricant and represents a vast improvement over previous results. At least tens of billions of cycles and demonstration of uniform endurance across many devices operating in parallel would be required for a device to be commercially viable. In a promising advancement for MEMS reliability, Texas Instruments was successful in producing light projectors consisting of an array of digital micromirror devices (DMDs) that function by switching between two orientation angles, corresponding to "on" and "off" positions. DMDs rely on both an advanced (vapor-phase) lubrication scheme (Henck, 1997) and the constant mechanical oscillation of the mirror yolk to prevent adhesion of the yolk's contact points to the landing pads. DMDs have been commercialized successfully, but their functionality does not rely on a sliding interface. Further advanced systems that incorporate many sliding interfaces will require lubrication schemes with a level of sophistication on par with the complexity of the device.

As these processing and operational barriers are overcome, and premature failure due to adhesion and wear are eliminated, issues of energy dissipation that continue to pose challenges for electronic devices will become equally important for M/NEMS. Frictional energy dissipation can be very low for individual nanoscale contacts, which consist of only tens or hundreds of atoms. But complex M/NEMS devices involve many of these single-asperity contacts operating in parallel (Romig et al., 2003). Using conservative values for the friction force for a single-asperity (1 nN to 100 nN) and for the sliding velocity (10^{-6} m/s to 10^{-2} m/s), the frictional energy dissipated by a single nanoscale contact can range from a femtowatt (1 fW = 10^{-15} W) to a few nanowatts (1 nW = 10^{-9} W). However, at more technologically relevant sliding velocities (i.e., > 1 mm/s), these losses are more than the total effective power dissipated by a single silicon transistor in an integrated complementary metal-oxide-semiconductor (CMOS) circuit. (At maximum allowed power densities, a modern transistor dissipates approximately one-tenth of a picowatt (1 pW = 10^{-12} W) [Lin and Banerjee, 2008].) Furthermore, nanoscale asperities are more densely packed than CMOS transistors, possibly increasing the total energy dissipated per area and ultimately lost as heat. M/NEMS employing components that rely on sliding contacts comprise thousands of nanoscale contacts per device. Assuming over a million devices operating in parallel (as in a micromirror array), with each device consisting of thousands of single-asperity contacts sliding at more than 1 mm/s, up to several watts of input power will be lost to friction alone. Hence, the ability to tune friction either by surface treatments or tailored substrate materials is critical for the future of complex M/NEMS technologies.

Toward this end, this chapter describes the tools and models used to understand the fundamental properties of friction at the nanoscale. Section 5.2 provides a description of the atomic force microscope (AFM) and explains how the AFM is used to measure friction and extract shear strength, work of adhesion, and other important properties. The goal is to provide a starting point for researchers interested in performing nanoscale friction measurements, using the most widely available instrumentation and contact models. Section 5.3 concludes with a brief discussion of new measurement techniques and modeling challenges important for the development of commercially viable devices.

5.2 FRICTION FORCE MICROSCOPY

The first measurement of nanoscale friction was performed by Mate, McClelland, Erlandsson, and Chiang in 1987 (Mate et al., 1987). They observed atomic-level stick-slip behavior on graphite: The atoms on the nanoscale apex of a tungsten AFM tip would stick and slip laterally along the surface with the spatial periodicity of the graphite (0001) lattice below. An example of the same type of stick-slip behavior is shown in Figure 5.2 for an AFM tip sliding on freshly cleaved sodium chloride, and atomic level stick-slip behavior is discussed in detail elsewhere (Morita et al., 1996). Binnig, Quate, and Gerber had invented the AFM over a year prior to Mate's

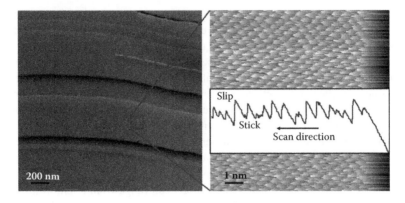

FIGURE 5.2 Left: 2 μm × 2 μm AFM deflection image of atomic terraces on the (001) face of a freshly cleaved sodium chloride crystal in ultrahigh vacuum. (The curvature is caused by drift in the piezoelectric scanner.) Right: A 10 nm × 10 nm stick-slip image and cross section resolving the atomic lattice on the terrace indicated to the left. (The lattice constant of NaCl is 0.564 nm.)

remarkable discovery (Binnig et al., 1986). They modified the scanning tunneling microscope (STM) (Binnig and Rohrer, 1982, 1983) by affixing a diamond probe to a small, compliant cantilever placed between an STM tip and the sample surface. This configuration permitted them to measure and to vary the forces between the probe tip and sample surface with sub-nanoNewton precision. Here, the tunneling current between the STM tip and the conductive Si cantilever was used to measure the displacement of the free end of the compliant cantilever and therefore the normal load applied to the sample. The spatial resolution of the instrument approached atomic-scale precision. The AFM thus enabled both force and topographical measurements on *insulating* and conducting surfaces, alike, with high resolution. In their pioneering nanoscale friction experiments, in which they also observed atomic-level stick-slip for the first time, Mate et al. (1987) used optical interferometry (instead of an STM approach) to sense the *torsion* of the cantilever and thereby measure *lateral* forces at the atomic level.

Later, the use of a laser beam position-sensitive detector (PSD) became the primary approach for sensing the minute angular deflections of the cantilever beam (Meyer and Amer, 1988; Meyer and Amer, 1990a; Alexander et al., 1989). As in the original STM-based AFM, the cantilever would bend in response to nanoscale forces on the probe tip attached at its free end. In the approach using a PSD, however, the change in angle of the cantilever translates to a change in the angle of reflection of a laser beam incident on the reflective backside of the cantilever. The position of the laser spot on the photodetector depends on this reflection angle, and the resulting "deflection" signal can be calibrated in units of force (as described in the following text) and used to determine the applied load in an experiment. Most measurements of nanoscale friction extend this optical-beam-deflection technique to measure the torsion of a cantilever beam in response to lateral forces. In fact, the two groups of Marti, Colchero, and Mlynek (Marti et al., 1990) and of Meyer and Amer (1990b) expanded upon this method to record the normal and lateral displacements of the laser beam *simultaneously*. (The normal and lateral signals correspond to the bending and torsional response of the cantilever, respectively.) The ability to record these two signals separately but simultaneously enables coincident bending and torsion measurements, such as the one shown for graphite in Figure 5.3. Today, this dual-force optical-beam-deflection method is the most common technique used in friction force microscopy (FFM). Many important FFM studies have relied on the instrumentation described above, and extensive reviews that together map out the history of measurement and experiments in nanotribology can be found elsewhere (Carpick and Salmeron, 1997a; Hähner and Spencer, 1998; Meyer et al., 1998; Persson, 2000; Gnecco et al., 2001; Krim, 2002; Colton, 2004; Szlufarska et al., 2008).

The AFM controls the position of the fixed end of the cantilever, which permits indirect control over the applied load in an experiment, with the load range and sensitivity depending on the spring constant of the cantilever. As a result, the AFM is prone to snap-in and pull-off (snap-out) instabilities, which occur at a tip–sample separation where the interaction force gradient matches the cantilever spring constant. Another highly valuable but less common scanning probe instrument used to investigate tribological properties is the interfacial force microscope (IFM) (Joyce and Houston, 1991). Through the use of a capacitive platform to sense cantilever deflection, the IFM enables direct control over the applied load and is therefore immune to the instabilities inherent in the AFM. The original purpose of the IFM was to investigate mechanical properties and adhesion by controlling tip–sample separation (Houston and Michalske, 1992); it was not explicitly designed for performing FFM measurements. Nonetheless, it has been used in several cases for friction measurements in the tensile regime, where AFM measurements can be challenging (Burns et al., 1999; Kiely and Houston, 1999; Kim and Houston, 2000a; Houston and Kim, 2002; Major et al., 2003; Major et al., 2006). However, quantitative analysis of friction measurements in an IFM is limited, because it is difficult to separate the torsional signal (due to friction) from the normal force signal and to calibrate each force independently.

The ability to measure normal and lateral signals simultaneously is essential for measuring nanoscale friction, as the properties of friction, including stick-slip behavior and the magnitude of friction forces, are typically load dependent. The linear load dependence observed at the macroscale arises from the micro- and nanoscale roughness of the contacting surfaces causing an increase in true contact area with increasing load: As the pressure on each protrusion (or *asperity*) increases, the contacting asperities deform and flatten (Greenwood and Williamson, 1966), and this elastic (and sometimes plastic) deformation reduces the overall gap between the surfaces and permits shorter protrusions to make contact with the counter-surface, increasing total contact area. Accordingly, the familiar (macroscale) coefficient of friction, which is taken to be equal to the slope of the friction–load curve, is not an intrinsic property of an interface, because it depends on surface geometry (e.g., roughness) and wear. Therefore, it is important to measure friction for individual nanoscale asperities in order to extract truly intrinsic quantities. While friction is load dependent even at the single-asperity level, the advantage of performing friction measurements at the nanoscale is the relative ease with which the contact geometry may be modeled

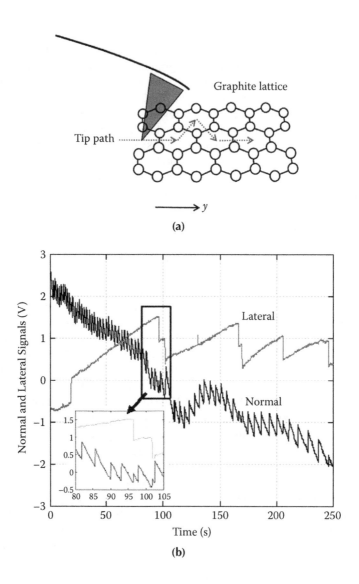

FIGURE 5.3 (a) Schematic representation of an AFM tip sliding along the graphite (0001) lattice, indicating a possible tip path, which can deviate from the direction (y in this depiction) along which the sample or tip base is scanned. (b) Stick-slip occurring as a function of time while the sample is moved vertically. The cantilever bends and releases, because its tilt angle (in this case, 22.5°) with respect to the sample couples z-motion into a tip-sample displacement along y. This forces the tip (silicon nitride, with $k_N = (0.022 \pm 0.001)$ N/m) to slide along the freshly cleaved graphite surface, resulting in stick-slip behavior. Most of the stick-slip events occur along y; however, jumps along x are also observed. This occurs when the tip slips sideways into a nearby honeycomb structure, as indicated in (a). The inset in (b) reveals the half-jumps in the normal signal that coincide with each of these lateral slip events.

and the ability to identify and avoid atomic-level wear by precise control of contact pressures.

In this regard, continuum mechanics has proved to be a useful tool for determining the load dependence of the tip–sample contact area. Contrary to macroscopic observations, Bowden and Tabor found that friction is proportional to the true area of contact (Bowden and Tabor, 1985). Consequently, if the area–load relation is known, a friction–load plot can reveal critical information about the interface, including the work of adhesion, pressure dependence, and wear properties. Continuum mechanics models can yield quantitative area–load relationships when the elastic properties of the two contacting materials are known. It should be noted, however, that the breakdown of continuum mechanics at the nanoscale is a critical topic of ongoing interest, and caution must be used when applying a continuum fit to nanoscale friction–load data (Luan and Robbins, 2005, 2006; Mo et al., 2009). This breakdown can occur when atoms are treated as discrete entities, as opposed to forming a smooth, continuous line of material that fills and defines the boundaries of an object. Measuring nanoscale contacts and forces with sub-nanoNewton precision gives us insight into potentially the most fundamental mechanisms of friction at all scales, but it also means that these data are more sensitive to microscopic laws. Fortunately, while the most appropriate analysis approach is not yet established nor fully understood for each experimental system, free parameters that emerge from continuum methods can be checked for consistency with separately calculated physical quantities, as discussed in detail in the following text.

Before describing the continuum mechanics models, it is important to discuss the apparatus and data collection procedure. Simultaneous measurement of normal and lateral signals in an AFM requires a four-quadrant PSD (instead of only two photodiodes required for a simple normal force measurement). Rastering of the probe tip over the sample is accomplished by actuating a piezoelectric (piezo) scanner in the xy plane, while controlling z-motion to maintain constant height or applied load. Figure 5.4 shows a basic schematic of an AFM, including the (now-standard) four-quadrant detector, xyz scanner with sample, and a rectangular cantilever and probe tip. The scanner designs vary depending on the instrument, with major differences being the decoupling of the z-piezo from the xy stage and the use of closed-loop scanners to ensure accurate displacements.

The deflection, or *normal*, signal, V_N, is obtained by subtracting the top (A and B) and bottom (C and D) PSD quadrants indicated in Figure 5.4:

$$V_N = (V_A + V_B) - (V_C + V_D) \tag{5.1}$$

Likewise, the difference between left and right quadrants gives the lateral signal, V_L:

$$V_L = (V_A + V_C) - (V_B + V_D) \tag{5.2}$$

Both the normal and lateral signals scale with the total, or *sum*, signal of all the quadrants. Dividing out the sum signal is sometimes performed but is unnecessary, as long as the normal and lateral force calibration factors account for any changes in total signal. In a nanoscale friction measurement, the applied load is typically

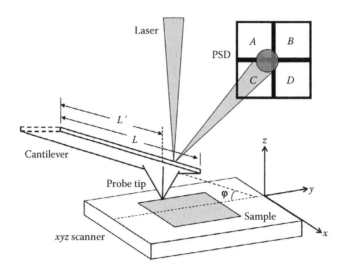

FIGURE 5.4 Schematic of an AFM employing the typical optical-beam-deflection technique. Components include a cantilever and probe, four-quadrant PSD, *xyz* scanner, and sample. Both the full length, L, of the cantilever and distance between the tip and cantilever's base, L', are indicated. The angle, ϕ, corresponds to the incline of the cantilever's long axis with respect to the *xy* plane of the scanner.

ramped line-by-line in the AFM image, as the tip rasters over the sample in the *xy* plane. By scanning forward and backward in the +*x* (*trace*) and –*x* (*retrace*) directions, respectively, a friction loop is recorded at each applied load. Figure 5.5a shows a basic friction loop. To remove contributions from the lateral signal offset, the friction force is obtained by calculating the half-width of the loop (as in Figure 5.5a), equal to half of the difference between the trace and retrace signals. Friction is then calibrated in units of force, using methods described in the following text. The load is varied either by adjusting the *z*-piezo position or by directly controlling the deflection set point in a feedback loop with the *z*-piezo. In either case, each load will be applied to different locations on the sample, as significant displacement of the tip along the *y*-direction occurs due to the tilt of the cantilever. This effect (exemplified in Figure 5.3b) was first observed by Watson et al. (2004) and can be countered with a scanner displacement in the direction opposite to the lateral tip displacement, as described elsewhere (Cannara et al., 2005).

As shown for the stick-slip measurement on muscovite mica in Figure 5.5b, *x*-scan sizes as low as a few nanometers may be used, depending on the length of the initial sticking portion of the friction loop (e.g., line ab in Figure 5.5a and between 0 nm and 0.75 nm in the upper trace of Figure 5.5b). The applied load typically ranges from a maximum of one to three times the adhesive force to a minimum value at pull-off (where the applied load is equal and opposite to the adhesive force). The value of the maximum applied load is selected to avoid wear of the tip or sample surface or to obtain sufficient data for a continuum fit. The resulting data consist of a set of three simultaneously recorded images: (1) the raw deflection (normal) signal, (2) the lateral signal in the trace direction, and (3) the lateral signal in the retrace direction.

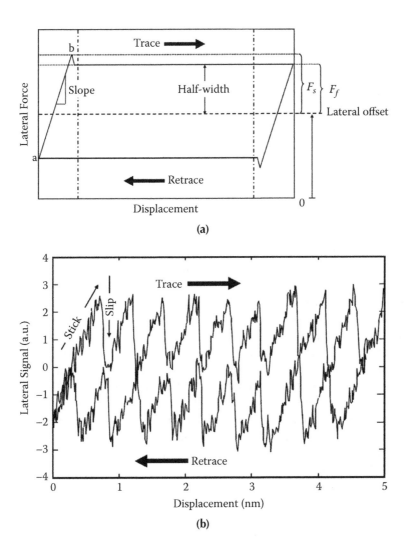

FIGURE 5.5 (a) Schematic friction loop consisting of a lateral offset (due to optical cross-talk and/or the geometry of the surface) and lateral forces measured in both the trace and retrace directions. A trace (or retrace) scan begins with the sticking portion indicated by line ab, whose slope corresponds to the total lateral stiffness of the cantilever-tip-contact system. Once the tip overcomes the lateral sticking force, F_S, it slides at the dynamic friction force, F_L, which can be determined from the half-width of the friction loop bounded by the two vertical dashed-dotted lines. (b) Stick-slip friction on freshly cleaved muscovite mica in ambient air. The stick and slip events repeat with the spatial periodicity of the mica lattice (≈ 0.529 nm). The top and bottom traces correspond to scanning the tip along forward (to the right) and reverse (to the left) directions, respectively.

The normal signal image contains an out-of-contact portion, which indicates the normal signal offset. This offset is subtracted from a line-by-line average of the image, which can then be calibrated in nN, as described in the next paragraph. The friction force is calculated from the difference between images (2) and (3) to obtain the half-width, as discussed previously. To remove contributions from the sticking portion (line ab in Figure 5.5a) of the data, only the region between the two vertical dashed-dotted lines in Figure 5.5a is included in the analysis.

The normal force, F_N, is calibrated with the normal spring constant of the cantilever and the deflection sensitivity of the optical system. While nominal values for the normal spring constant, k_N, are often supplied by the manufacturer, they can differ significantly from their actual values. Sader and coauthors have developed rapid, quantitative methods for determining the spring constants of cantilevers of various shapes (e.g., rectangular, triangular, etc.), using the plan view dimensions of a cantilever and its resonance properties in air (Sader, 1999; Green et al., 2004; Sader et al., 2005). Both the resonance frequency and quality factor may be determined by oscillating the cantilever in the AFM and measuring the response at different frequencies. This measurement must be performed in air (or in a medium with a sufficient viscosity) to accommodate the use of a hydrodynamic damping function in the calculation of k_N. Other reliable methods obtain resonance properties and spring constants based on the shift in fundamental-mode frequency due to an added mass (Cleveland et al., 1993; Green et al., 2004) or from the thermally induced oscillations (or *thermal noise*) of the cantilever (Hutter and Bechhoefer, 1993; Butt and Jaschke, 1995; Burnham et al., 2003). Alternatively, the spring constant of individual cantilevers can be determined directly by pressing against a reference cantilever of precisely known stiffness (Gibson et al., 1996; Tortonese and Kirk, 1997; Gates and Reitsma, 2007).

The spring constant of the *full* length of the cantilever is given by k_N, but the relevant spring is a function of the location along the cantilever's long axis where the load is actually applied in the experiment. Hence, if a calibration method determines k_N, it is necessary to measure the location of the tip and calculate an effective spring constant, k'_N according to the following equation:

$$k'_N = k_N \left(\frac{L}{L'} \right)^3 \tag{5.3}$$

where L is the full length of the cantilever, and L' is the distance from the base of the cantilever to the tip location along the cantilever's long axis. The deflection sensitivity, S_N, is the measured change in normal signal per change in z-displacement, when the tip is pressed against a *rigid* sample, such as silicon. (To measure S_N accurately, no sample or tip deformation may occur.) The normal force calibration factor, C_N, is then given by:

$$C_N = \frac{k'_N}{S_N \cos^2 \varphi} \tag{5.4}$$

The cosine-squared term accounts for the tilt angle, ϕ, of the cantilever with respect to the sample plane (Figure 5.4) to obtain the component of the force that is normal to the surface and the projection of the z-displacement perpendicular to the cantilever in the yz plane. Further consideration of the impact of cantilever tilt on force calibration yields a more complicated relationship, depending on the tip location, geometry and tip–sample interaction, as studied elsewhere (Stiernstedt et al., 2005; Edwards et al., 2008). Once the calibration factor is known for a specific experiment, accounting for tip location and cantilever tilt, the applied and total (applied plus adhesive) loads in the experiment can be calculated by multiplying the deflection signal (minus its offset) by C_N.

Calculating the lateral force calibration factor, C_L, is more complicated than the normal force calibration procedure. However, several approaches have been developed that are relatively straightforward (Munz, 2010). Each method falls along a spectrum that varies from the direct experimental determination of C_L to analytical or numerical methods, and the appropriate technique for a given experiment is chosen based on its accuracy, ease of use, and whether it is acceptable to risk damage to the tip. The most widely applicable techniques may be grouped under two headings defined here as *relative* and *direct*. Relative methods require prior knowledge of either k_N or C_N to determine C_L, and direct methods do not. For example, the *wedge method* (Ogletree et al., 1996; Varenberg et al., 2003; Tocha et al., 2006), originally developed by Ogletree et al., is an indirect method, as its output is the ratio between lateral and normal force calibration factors, and a linear friction–load relationship is assumed in the derivation of this ratio. The wedge method is a well-established lateral force calibration technique and a standard to which newer methods are often compared. But it should be noted that the technique itself is a friction–load experiment performed on a two-sloped calibration sample (e.g., 1-μm to 2-μm-high (111) silicon facets) and thus incurs some risk of contamination and permanent damage to the tip. Performing the calibration with a separate, identical probe is possible but adds error to the measurement.

There are at least three calibration procedures that do not require sliding the tip over the surface of a calibration sample. The *test probe method* (Cannara et al., 2006; Chung et al., 2010) is the lateral analog to the normal force calibration procedure. It is therefore a direct method but requires knowledge of the cantilever's torsional spring constant, k_L, which can be obtained via the Sader torsional method (Green et al., 2004), and an understanding of the in-plane deformation of the cantilever (Sader and Green, 2004). In the test probe method, a spherical probe is attached to the free end of the cantilever, and a lateral force–displacement curve is obtained by pushing the probe against a rigid, vertical wall. The lateral deflection sensitivity (calculated from the force curve), along with k_L and the size of the sphere, yield the lateral force calibration factor, C_L. Depending on the experiment, C_L may then be used to calculate the calibration factor for a cantilever-*tip* combination with a similar cantilever geometry, or for the same cantilever with a sphere glued to the end postexperiment. This is valid, as long as the lateral stiffness of the tip shaft is large compared with the torsional spring constant. Alternatively, if the sphere itself is used both for calibration and experiment (instead of a tip), the respective contact points are the side and apex of the sphere, respectively, and tip stiffness can be neglected due to the low aspect ratio of the spherical probe. All variations of the test probe method avoid contacting the probe apex to a calibration sample, which is advantageous for probes with sensitive or unique end structures or coatings. If there is contamination or wear, it is then certain that it occurred during the experiment and not during the calibration procedure.

Direct calibration of lateral forces without prior knowledge of k_L is achieved using the diamagnetic lateral force calibrator (D-LFC), which consists of a static friction measurement on a pyrolitic graphite sample levitating over four permanent magnets (Li et al., 2006). The tip makes contact with the sample, but no sliding occurs when the procedure is performed properly. Moreover, the freshly cleaved graphite surface is clean and inert, minimizing the possibility of tip damage. Unlike the test probe method, k_L is not required. Although the cantilever's spring constants remain unknown, the spring constant of the levitation system is characterized by a ring-down measurement, which yields the frequency and quality factor for the levitating sample as it oscillates about the minimum of the magnetic field. The main premise of the D-LFC method is that the levitation spring constant is so much lower than the contact spring and the torsional spring constant of the cantilever that the lateral displacement of the tip is negligible in comparison with the lateral displacement of the sample. As a result, the static friction data represent the torsional response of the cantilever to a force approximately equal to the scan size times the levitation spring constant. Much like the test probe and normal force calibration equations, the slope of the lateral signal as a function of scanner displacement (typically in V/nm) can be divided by the levitation spring constant to obtain C_L. This direct calibration of C_L provides a clever way of avoiding the limitations suffered by other static methods (Cain et al., 2001).

A third direct lateral force calibration technique determines C_L by exerting a torque at a known distance from the long axis of the cantilever near the tip location (Feiler et al., 2000). Reitsma et al. improved upon this *pivot method* by performing the measurement at multiple pivot points. Their approach requires either a special "hammerhead" probe (Reitsma, 2007) or a wide enough cantilever beam such that multiple distinct locations near the free end of the cantilever may be loaded (Chung and Reitsma, 2007). The result is a simultaneous measurement of the torsional spring constant of the cantilever beam and the lateral deflection sensitivity of the optical system. The benefits of this calibration technique include the lack of sliding or contacting the tip to a calibration surface and the absence of any lateral scanning, which would require a calculation of in-plane bending (Sader and Green, 2004). As with all methods, the effect of cantilever tilt on force calibration remains important, as tilt can have a significant impact on both normal and lateral force calibration factors alike (Stiernstedt et al., 2005; Edwards et al., 2008). The pivot method may also be used to determine k_L.

A note about probe materials: Probe tips may consist of a variety of conducting or insulating materials. While silicon or silicon nitride can be etched from a single crystal, variations in tip materials are typically achieved by coating silicon probes (e.g., with titanium nitride, silicon nitride, tungsten carbide, platinum, etc.) Much effort, however, has gone into producing monolithic structures composed of hard materials, including diamond, polycrystalline diamond, and diamond-like carbon (DLC) (Sullivan et al., 2001; Sekaric et al., 2002; Olivero et al., 2006; Peiner et al., 2007; Luo et al., 2007). Similar to the silicon or silicon nitride cantilevers, in a monolithic structure, the entire cantilever–probe combination is made of one continuous structural material. *Unlike* silicon or silicon nitride, amorphous or polycrystalline diamond is deposited instead of etched. To date, no monolithic single-crystal diamond cantilever–probe combination has been reported; however, Obraztsov et al. (2010) recently developed a technique for producing individual diamond tips on silicon cantilevers by a series of chemical vapor deposition, oxidation, and silicon etch steps. To obtain *amorphous* diamond tips, Bhaskaran et al. (2010) deposited DLC into molds in silicon and fabricated silicon cantilevers with DLC tips. It is also possible to coat tips with DLC and various hydrocarbon materials by electron beam–induced deposition (EBID) in an electron microscope (Schwarz et al., 1997). The scanning electron microscope and the transmission electron microscope (TEM) are useful for both EBID and tip-shape measurements (Derose and Revel, 1997; Kopycinska-Müller et al., 2006). In particular, the TEM is indispensable for high-precision measurements of tip geometry (e.g., radius of curvature), coating thickness, and local atomic structure. As will be evident in the data analysis discussion to follow, an accurate tip radius measurement is critical for extracting quantitative information from friction–load curves.

At least two important quantities are obtained from nanoscale friction measurements: the shear strength (τ) and the work of adhesion (w). As described previously, τ is the intrinsic resistance to sliding. The work of adhesion is the energy per unit area that must be supplied to pull two surfaces apart. It can be quantified in terms of surface energies:

$$w = \gamma_1 + \gamma_2 - \gamma_{12} \qquad (5.5)$$

where γ_1 and γ_2 correspond to the energies of formation of the two surfaces, and γ_{12} corresponds to the energy required to create the interface. For identical surfaces, γ_{12} is equal to zero; for dissimilar materials, elastic deformation or the off-equilibrium positioning of interfacial atoms leads to additional (potential) energy stored in the contact, which is then regained upon separation. In general, w will affect the magnitude and load dependence of the tip–sample contact area and, consequently, the friction force. However, to first order, τ is independent of w.

The shear strength can depend upon parameters such as pressure, temperature, and velocity (Briscoe and Evans, 1982), to name a few. As mentioned previously, the friction force, F_L, is proportional to the contact area, A, with τ the constant of proportionality (Bowden and Tabor, 1985):

$$F_L = \tau A \qquad (5.6)$$

The dependence of τ on pressure, P, is often assumed to be linear:

$$\tau = \tau_0 + \zeta P \qquad (5.7)$$

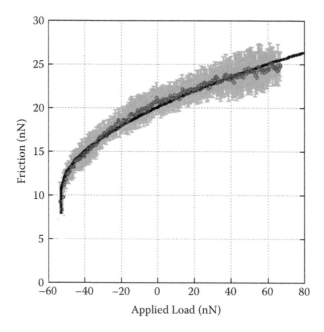

FIGURE 5.6 Friction plotted as a function of applied normal load (open circles); error bars are the standard deviations from the mean indicating the curve-to-curve variation. The data are fit by the COS transition model (solid black line), representing the dependence of contact area on applied load.

To first order, however, it can be assumed that the shear strength has no pressure dependence ($\zeta = 0$), and the shape of the friction–load curve is determined solely by the change in A with applied load. Figure 5.6 provides an example of a friction–load curve for a hydrocarbon-coated tip sliding on microcrystalline diamond in dry nitrogen (150 nm tip radius, 20 nm scan size). The solid line is a curve fit assuming constant shear strength and using the continuum methods described in the following text.

The area–load relationship is a function of tip radius, R, the elastic properties of the two materials, and the magnitude and range of their interaction (i.e., w). These physical properties determine the location of the interface along a continuous theoretical spectrum bounded by two continuum mechanical extremes (Greenwood, 1997). Specifically, the physical properties of the interface determine the fraction of elastic deformation that occurs due to adhesion of the unloaded interface with respect to the overall range of interaction between the two surfaces. This ratio is called Tabor's parameter and is given by (Tabor, 1977):

$$\mu_T = \left(\frac{16Rw^2}{9K^2 z_0^3} \right)^{1/3}$$

(5.8)

where z_0 is the equilibrium separation for a Lennard-Jones interaction potential, and

$$K = \frac{4}{3}\left(\frac{1-v_{tip}^2}{E_{tip}} + \frac{1-v_{sample}^2}{E_{sample}}\right)^{-1}$$

is the combined (or *reduced*) elastic modulus, given by the Poisson's ratios, v, and Young's moduli, E, of the tip and sample. At one end of the spectrum, the contact deformation is significant compared with the range of interaction (z_0), and μ_T is also large. In this case, the area–load relationship is described by Johnson-Kendall-Roberts (JKR) theory (Johnson et al., 1971). At the other end of the spectrum, the deformation is small compared with z_0. Thus, μ_T is small, and the contact follows Derjaguin-Muller-Toporov (DMT) theory (Derjaguin et al., 1975). Intermediate regimes exist between these two extremes, and the corresponding transition from JKR to DMT has been described analytically by Carpick, Ogletree and Salmeron (COS) (1997b) and given its physical basis by Schwarz (2003). The generalized transition, or *COS*, equations have made the JKR-DMT analysis easily applicable to nanoscale friction measurements (Grierson et al., 2005).

In general, the following COS equation describes the dependence of the contact area on normal load, F_N, for a sphere(or paraboloid)-on-flat geometry:

$$A = A_0 \left(\frac{\alpha + \sqrt{1 - F_N / F_C}}{1+\alpha}\right)^{4/3} \tag{5.9}$$

where A_0 is the contact area at zero applied load, and F_C is the minimum, or *critical*, applied load, corresponding to an instability in the tensile regime of the area–load curve ($F_C < 0$). While F_C is typically equal to the pull-off force in an experiment, pull-off can also occur prematurely, or *early*, (i.e., for $F_N > F_C$), in which case the magnitude of the measured pull-off force is less than the true adhesive force, $|F_C|$. The dimensionless fitting parameter, α, identifies the mechanical nature of the contact, and its physical origin can be traced back to Tabor's parameter. In fact, both A_0 and F_C depend on α, as they are directly related to the deformability of the contact and the range of interaction. In addition to identifying the location of the contact along the JKR-DMT spectrum, the value of α can be used to determine whether the use of continuum mechanics models is appropriate for the nanoscale interface; or conversely, whether the continuum approach breaks down at this scale. α is connected to μ_T via Maugis's parameter, λ, a simplified version of μ_T obtained using a square-well potential. In the COS equations, α is related to λ numerically, and this leads to the following approximation for the experimentally determined μ_T:

$$\mu_T(\alpha) \approx -0.7986 \ln(1 - 1.02\alpha) \tag{5.10}$$

If this experimental value is significantly different from a separate estimation of μ_T that is based on physically reasonable values for z_0 and K (as described in detail elsewhere [Grierson et al., 2005]), then it is possible that this continuum approach is inappropriate for the interface, or that the assumption of a constant shear strength is false. If w is known separately, comparison of Equations (5.8) and (5.10) can be performed without calibrating the normal and lateral forces, as α does not depend on C_N or C_L.

Once the calibration and fitting procedures are complete, w and τ_0 are calculated from α, R, and K, according to the following relations:

$$w = \frac{-F_C}{\hat{F}_C(\alpha)\pi R} \tag{5.11}$$

$$\tau_0 = \frac{F_{f,0}}{A_0} \tag{5.12}$$

where

$$A_0 = \left(\frac{\pi w R^2}{K} \right)^{2/3} \hat{A}_0(\alpha) \tag{5.13}$$

The dimensionless, α-dependent terms, $\hat{A}_0$ and $\hat{F}_c$ emerge from the fit, as outlined in the original COS paper (Carpick et al., 1997b). In this procedure, it is typical to use bulk values for the elastic constants. While this approach has produced reasonable results in certain cases, it is important to consider that it may fail in others. Moreover, while a sphere-on-flat is often a good approximation for the interfacial geometry, tip shape has a strong impact on the area–load relation (Carpick et al., 1996). In addition to the use of electron microscopy to characterize tip shape (Kopycinska-Müller et al., 2006), the blind reconstruction scheme of Villarubia allows tip shape to be extracted from topographic images of sharp nanoscale features (Villarrubia, 1996; Villarrubia, 1997; Dongmo et al., 2000). If early pull-off occurs, it may be necessary to consider in the analysis the effect of contact area reduction, or *microslip*, caused by the applied shear stress (Johnson, 1997; Unertl, 1999).

Up until this point, it has been assumed that the shear strength is independent of load. This assumption precludes an unbiased understanding of the intrinsic frictional response at an interface. As discussed above, if a continuum fit is applied directly to friction–load measurements, a discrepancy between calculated and estimated Tabor's parameters may indicate that this assumption is incorrect (though it may also indicate that continuum methods are altogether inapplicable). There are at least two ways to measure contact area directly and thereby resolve this dilemma: (1) lateral stiffness and (2) electrical conductance measurements. Conductance increases with contact area and thus varies with load (Pethica and Tabor, 1979), permitting the independent measurement of area–load relationships in an AFM. This has been demonstrated for a tungsten carbide tip on boron-doped diamond (Enachescu et al., 1998). However, the requirement that both materials be conducting restricts the use of contact conductance measurements to a small subset of friction–load experiments.

On the other hand, lateral stiffness measurements are, in principle, useful for any interface. The measurement is accomplished by a lock-in technique described elsewhere (Carpick et al., 1997c; Lantz et al., 1997a).

In an AFM, the total lateral stiffness is given by the torsional spring constant of the cantilever, the lateral stiffness (bending) of the tip shaft, and the lateral contact stiffness:

$$k_{L,total}^{-1} = k_{L,cantilever}^{-1} + k_{L,tip}^{-1} + k_{L,contact*}^{-1} \qquad (5.14)$$

The contribution from $k_{L,tip}$ is important to consider for tip shafts with high aspect ratios. Furthermore, it has been shown that the low contact area of sharp tips can lead to contact stiffnesses comparable to the torsional spring constant of the cantilever, and thus have a significant impact on the calculation of lateral forces (Lantz et al., 1997a; Piétrement et al., 1999). The lateral contact stiffness arises from the gradient of the interaction force between atoms on the tip and sample. These individual springs act in parallel. As a result, their contributions are summed together, and their total effect is proportional to the contact area. Hence, the contact stiffness is generally a function of the applied load. Lateral stiffness measurements yield $k_{L,total}$; therefore, $k_{L,cantilever}$ and $k_{L,tip}$ must be determined separately in order to extract $k_{L,contact}$. Furthermore, for a sphere-on-flat geometry, the equation relating lateral contact stiffness to contact area,

$$k_{L,contact} = 8G^* \sqrt{\frac{A}{\pi}} \qquad (5.15)$$

depends on prior knowledge of the combined shear modulus,

$$G^* = \left(\frac{2 - v_{tip}}{G_{tip}} + \frac{2 - v_{sample}}{G_{sample}} \right)^{-1}$$

where G is the shear modulus of the tip or sample. For their measurements, Lantz et al. (1997b) used bulk values for G^* and demonstrated good agreement in comparison with continuum theory. If the relevant elastic moduli are known, lateral stiffness versus load measurements yield the area–load relation:

$$A(F_N) = \frac{\pi k_{L,contact}^2 (F_N)}{64 G^{*2}} \qquad (5.16)$$

$\tau(F_N)$ is obtained by combining Equations (5.6) and (5.16). In this way, any pressure (or load) dependence of the shear strength is revealed. Otherwise, dividing the two experimentally obtained quantities, F_L and $k_{L,contact}^2$, yields the simple proportionality:

$$\frac{F_L}{k_{L,contact}^2} = \frac{\pi \tau_0}{64 G^{*2}} \qquad (5.17)$$

Multiple authors have determined lateral stiffness and contact area as a function of applied load for various materials and conditions, using this technique (Carpick et al., 1997c; Lantz et al., 1997a; Lantz et al., 1997b; Piétrement et al., 2001a, 2001b; Wahl et al., 1998). Piétrement et al. further developed the lateral stiffness measurement technique to avoid the use of elastic constants in the contact area calculation completely (2001b). Alternatively, contact stiffness measurements may be used to calculate local elastic properties (Reinstädtler et al., 2005; Hurley and Turner, 2007; Hurley, 2010). Altogether, lateral stiffness measurements have provided independent confirmation of Bowden and Tabor's hypothesis that friction is proportional to contact area.

5.3   OUTLOOK FOR NANOSCALE DEVICES

It is critical to emphasize that the analysis presented in Section 5.2 is based on the assumption that continuum methods accurately model nanoscale contacts. Simulations of atomistic behavior have shown that deviations from a continuum approach can occur, depending on the atomic-level geometry of the contact. In addition to the general shape (e.g., parabolic, flat punch, etc.) or the nanoscale roughness of the contacting bodies, the relative arrangement of atoms at the interface can have a strong impact on nanoscale friction and area–load relations. This is underscored by the existence of atomic-level stick-slip. The following atomic-scale geometries have been simulated: commensurate (where the two surfaces have coincident or perfectly matched periodicities or some degree of overlap), incommensurate (where they exhibit no overlap), or amorphous (where at least one surface consists of randomly arranged atoms). In their simulations, Müser et al. found that rigid, incommensurate surfaces can interlock with each other only in the presence of mobile atoms or a diffusing film (or fluid) at their interface (Müser and Robbins, 2000; Müser et al., 2001). This leads to increased static friction. If the two materials are incommensurate, but at least one of them is sufficiently compliant, local stresses can cause the atoms at their interface to move into interlocking positions. Surprisingly, the same study by Müser et al. found that, although the diffusing atoms in the film are mobile, they fail to dislodge a commensurate interface. In fact, based on their findings, a thin fluid layer can cause interlocking between commensurate and incommensurate surfaces alike. Separately, Kim and Hurtado have studied length scale effects: Based on a dislocation-assisted slip model, they predict that the shear strength should increase as the tip radius (or object size) decreases (Kim and Hurtado, 2000b; Hurtado and Kim, 1999a, 1999b). The implications for M/NEMS devices are profound.

Returning to issues of M/NEMS reliability, it should be noted that friction and wear are inextricably related. High shear stress caused by large lateral forces (friction) leads to wear. If the interaction between atoms at an interface is relatively low and the corrugated surface potential shallow, then sliding can be easy for atomically smooth surfaces, and the bonds between the atoms and their substrates are less likely to stretch and break. Observations of macroscale phenomena support this interpretation; however, aside from very recent work (Gotsmann and Lantz, 2008; Lantz et al., 2009; Bhaskaran et al., 2010), few studies have focused directly on the impact of shear stress on wear at the *atomic* level. As with friction, it is clear from these studies that macroscopic models do not accurately represent nanoscale observations, but a full understanding of atomic-scale wear has not yet emerged. The combination of electron microscopy and AFM or similar tools to observe mechanical deformation (both elastic and plastic) and materials transformations is an enabling technology in this respect (Erts et al., 2002; Cumings and Zettl, 2000). In particular, the TEM has enabled the in situ observation of atomic-level phenomena, including, for example, the mechanical annealing of defects in nanoscale structures (Shan et al., 2008) and the formation of single-atom-thick gold wires (Takai et al., 2001). Further use of combined AFM/TEM instruments for tribological measurements would enhance the understanding of both frictional energy dissipation and atomic-scale wear.

While it is important to understand the impact of different parameters and conditions to improve tribological behavior, it is also interesting to approach new discoveries as a means to harness the frictional properties of a material for high-precision manipulation and control. It is already known that nanoscale friction may be actively tuned (or switched off and on) either electronically (Park et al., 2006, 2007) or by mechanical actuation (Socoliuc et al., 2006; Gnecco et al., 2009; Lantz et al., 2009). Furthermore, atomic-level stick-slip can provide a ratcheting mechanism useful for high-precision positioning, with each step equal to an integer multiple of the spacing between surface atoms or lattice sites. In this respect, the scientific and technological knowledge gained from nanotribological measurements has the potential for advancements beyond improving the performance of M/NEMS devices.

REFERENCES

Alexander, S., L. Hellemans, 0. Marti, J Schneir, V. Elings, P. K. Hansma, M. Longmire, and J. Gurley. 1989. An atomic-resolution atomic-force microscope implemented using an optical lever. *J. Appl. Phys.* 65: 164–167.

Asay, D. B., M. T. Dugger, J. A. Ohlhausen, and S. H. Kim. 2008a. Macro- to nanoscale wear prevention via molecular adsorption. *Langmuir* 24: 155–159.

Asay, D. B., M. T. Dugger, and S. H. Kim. 2008b. In-situ vapor-phase lubrication of MEMS. *Tribol. Lett.* 29: 67–74.

Bhaskaran, H., B. Gotsmann, A. Sebastian, U. Drechsler, M. A. Lantz, M. Despont, P. Jaroenapibal, R. W. Carpick, Y. Chen, and K. Sridharan. 2010. Ultralow nanoscale wear through atom-by-atom attrition in silicon-containing diamond-like carbon. *Nature Nanotechnol.* 5: 181–185.

Binnig, G., C. F. Quate, and C. Gerber. 1986. Atomic force microscope. *Phys. Rev. Lett.* 56: 930–933.

Binnig, G., and H. Rohrer. 1982. Scanning tunneling microscope, U. S. Patent, 4,343,993 (3 pages).

Binnig, G., and H. Rohrer. 1983. Surface imaging by scanning tunneling microscopy. *Ultramicroscopy* 11: 157–160.

Bowden, F. P., and D. Tabor. 1985. *The Friction and Lubrication of Solids.* Oxford, UK: Clarendon Press.

Briscoe, B. J., and D. C. B. Evans. 1982. The shear properties of Langmuir-Blodgett layers. *Proc. R. Soc. Lond. A* 380: 389–407.

Burnham, N. A., X. Chen, C. S. Hodges, G. A. Matei, E. J. Thoreson, C. J. Roberts, M. C. Davies, and S. J. B. Tendler. 2003. Comparison of calibration methods for atomic force microscopy cantilevers. *Nanotechnology* 14: 1–6.

Burns, A. R., J. E. Houston, R. W. Carpick, and T. A. Michalske. 1999. Friction and molecular deformation in the tensile regime. *Phys. Rev. Lett.* 82: 1181–1184.

Butt, H.-J., and M. Jaschke. 1995. Calculation of thermal noise in atomic force microscopy. *Nanotechnology* 6: 1–7.

Cain, R. G., M. G. Reitsma, S. Biggs, and N. W. Page. 2001. Quantitative comparison of three calibration techniques for the lateral force microscope. *Rev. Sci. Instrum.* 72: 3304–3312.

Cannara, R. J., M. J. Brukman, and R. W. Carpick. 2005. Cantilever tilt compensation for variable-load atomic force microscopy. *Rev. Sci. Instrum.* 76: 053706.

Cannara, R. J., M. Eglin, and R. W. Carpick. 2006. Lateral force calibration in atomic force microscopy: A new lateral force calibration method and general guidelines for optimization. *Rev. Sci. Instrum.*, 77: 053701.

Carpick, R. W., N. Agraït, D. F. Ogletree, and M. Salmeron. 1996. Measurement of interfacial shear (friction) with an ultrahigh vacuum atomic force microscope. *J. Vac. Sci. Technol. B* 14: 1289–1295.

Carpick, R. W., and M. Salmeron. 1997a. Scratching the surface: Fundamental investigations of tribology with atomic force microscopy. *Chem. Rev.* 97: 1163–1194.

Carpick, R. W., D. F. Ogletree, and M. Salmeron. 1997b. A general equation for fitting contact area and friction vs. load measurements. *J. Coll. Interf. Sci.* 211: 395–400.

Carpick, R. W., D. F. Ogletree, and M. Salmeron. 1997c. Lateral stiffness: A new nanomechanical measurement for the determination of shear strengths with friction force microscopy. *Appl. Phys. Lett.* 70: 1548–1550.

Chung, K.-H., J. R. Pratt, and M. G. Reitsma. 2010. Lateral force calibration: Accurate procedures for colloidal probe friction measurements in atomic force microscopy. *Langmuir* 26: 1386–1394.

Chung, K.-H., and M. G. Reitsma. 2007. Note: Lateral force microscope calibration using multiple location pivot loading of rectangular cantilevers. *Rev. Sci. Instrum.* 81: 026104.

Cleveland, J. P., S. Manne, D. Bocek, and P. K. Hansma. 1993. A nondestructive method for determining the spring constant of cantilevers for scanning probe microscopy. *Rev. Sci. Instrum.* 64: 403–405.

Colton, R. J. 2004. Nanoscale measurements and manipulation. *J. Vac. Sci. Technol. B* 22: 1609–1635.

Cumings, J., and A. Zettl. 2000. Low-friction nanoscale linear bearing realized from multiwall carbon nanotubes. *Science* 289: 602–604.

Derjaguin, B. V., V. M. Muller, and Y. P. Toporov. 1975. Effect of contact deformations and the adhesion of particles, *J. Colloid Interf. Sci.,* 53, pp. 314–326.

Derose, J. A., and J.-P. Revel. 1997. Examination of atomic (scanning) force microscopy probe tips with the electron microscope. *Microsc. Microanal.* 3: 203–213.

Dongmo, L. S., J. S. Villarrubia, S. N. Jones, T. B. Renegar, M. T. Postek, and J. F. Song. 2000. Experimental test of blind tip reconstruction for scanning probe microscopy. *Ultramicroscopy* 85: 141–153.

Edwards, S. A., W. A. Ducker, and J. E. Sader. 2008. Influence of atomic force microscope cantilever tilt and induced torque on force measurements. *J. Appl. Phys.* 103: 064513.

Enachescu, M., R. J. A. van den Oetelaar, R. W. Carpick, D. F. Ogletree, C. F. J. Flipse, and M. Salmeron. 1998. Atomic force microscopy study of an ideally hard contact: The diamond(111)/tungsten carbide interface. *Phys. Rev. Lett.* 81: 1877–1880.

Erts, D., A. Lõhmus, R. Lõhmus, H. Olin, A. V. Pokropivny, L. Ryen, and K. Svensson. 2002. Force interactions and adhesion of gold contacts using a combined atomic force microscope and transmission electron microscope. *Appl. Surf. Sci.* 188: 460–466.

Feiler, A., P. Attard, and I. Larson. 2000. Calibration of the torsional spring constant and the lateral photodiode response of frictional force microscopes. *Rev. Sci. Instrum.* 71: 2746–2750.

Fushinobu, K., L. M. Phinney, and N. C. Tien. 1996. Ultrashort-pulse laser heating of silicon to reduce microstructure adhesion. *Int. J. Heat & Mass Transfer* 39: 3181–3186.

Gates, R. S., and M. G. Reitsma. 2007. Precise atomic force microscope spring constant calibration using a reference cantilever array. *Rev. Sci. Instrum.* 78: 086101.

Gibson, C. T., G. S. Watson, and S. Myhra. 1996. Determination of the spring constants of probes for force microscopy/spectroscopy. *Nanotechnology* 7: 259–262.

Gnecco, E., A. Socoliuc, S. Maier, J. Gessler, T. Glatzel, A. Baratoff, and E. Meyer. 2009. Dynamic superlubricity on insulating and conductive surfaces in ultra-high vacuum and ambient environment. *Nanotechnology* 20: 025501.

Gnecco, E., R. Bennewitz, T. Gyalog, and E. Meyer. 2001. Friction experiments on the nanometre scale. *J. Phys.: Condens. Matter* 13: R619–R642.

Gotsmann, B., and M. A. Lantz. 2008. Atomistic wear in a single asperity sliding contact. *Phys. Rev. Lett.* 101: 125501.

Green, C. P., H. Lioe, J. P. Cleveland, R. Proksch, P. Mulvaney, and J. E. Sader. 2004. Normal and torsional spring constants of atomic force microscope cantilevers. *Rev. Sci. Instrum.* 75: 1988–1996.

Greenwood, J. A. 1997. Adhesion by elastic spheres. *Proc. R. Soc. Lond. A* 453: 1277–1297.

Greenwood, J. A., and J. B. P. Williamson. 1966. Contact of nominally flat surfaces. *Proc. Roy. Soc. Lond. A* 295: 300–319.

Grierson, D. S., E. E. Flater, and R. W. Carpick. 2005. Accounting for the JKR-DMT transition in adhesion and friction measurements with atomic force microscopy. *J. Adhesion Sci. Technol.* 19: 291–311.

Guckel, H., J. J. Sniegowski, T. R. Christenson, and F. Raissi. 1990. The application of fine-grained, tensile polysilicon to mechanically resonant transducers. *Sens. Actuators A* 21: 346–351.

Hähner, G., and N. Spencer. 1998. Rubbing and scrubbing. *Phys. Today* 51: 22–27.

Henck, S. A. 1997. Lubrication of digital micromirror devices TM. *Tribol. Lett.* 3: 239–247.

Houston, J. E., and H. I. Kim. 2002. Adhesion, friction, and mechanical properties of functionalized alkanethiol self-assembled monolayers. *Acc. Chem. Res.* 35: 547–553.

Houston, J. E., and T. A. Michalske. 1992. The interfacial-force microscope. *Nature* 356: 266–267.

Hurley, D. C. 2010. Measuring mechanical properties on the nanoscale with contact resonance force microscopy methods. In *Scanning Probe Microscopy of Functional Materials: Nanoscale Imaging and Spectroscopy*, ed. S. Kalinin and A. Gruverman. Berlin: Springer-Verlag.

Hurley, D. C., and J. A. Turner. 2007. Measurement of Poisson's ratio with contact-resonance atomic force microscopy. *J. Appl. Phys.* 102: 033509.

Hurtado, J. A., and K.-S. Kim. 1999a. Scale effects in friction of single-asperity contacts. I. From concurrent slip to single-dislocation-assisted slip. *Proc. R. Soc. Lond. A* 455: 3363–3384.

Hurtado, J. A., and K.-S. Kim. 1999b. Scale effects in friction of single-asperity contacts. II. Multiple-dislocation-cooperated slip. *Proc. R. Soc. Lond. A* 455: 3385–3400.

Hutchings, I. M. 1992. *Tribology—Friction and Wear of Engineering Materials*. Boca Raton, FL: CRC Press.

Hutter, J. L., and J. Bechhoefer. 1993. Calibration of atomic force microscope tips. *Rev. Sci. Instrum.* 64: 1868–1873.

Johnson, K. L. 1997. Adhesion and friction between a smooth elastic spherical asperity and a plane surface. *Proc. R. Soc. Lond. A* 453: 163–179.

Johnson, K. L., K. Kendall, and A. D. Roberts. 1971. Surface energy and the contact of elastic solids. *Proc. R. Soc. Lond. A* 324: 301–313.

Jost, P. 1966. *Lubrication (tribology)—A report on the present position and industry's needs.* Department of Education and Science, H.M. Stationary Office, London.

Jost, P. 1976. Economic impact of tribology. *Proc. Mechanical Failures Prevention Group,* NBS Special Pub., 423, Gaithersburg, MD.

Joyce, S. A., and J. E. Houston. 1991. A new force sensor incorporating force-feedback control for interfacial force microscopy. *Rev. Sci. Instrum.* 62: 710–715.

Kiely, J. D., and J. E. Houston. 1999. Contact hysteresis and friction of alkanethiol self-assembled monolayers on gold. *Langmuir* 15: 4513–4519.

Kim, H. I., and J. E. Houston. 2000a. Separating mechanical and chemical contributions to molecular-level friction. *J. Am. Chem. Soc.* 122: 12045–12046.

Kim, K.-S., and J. A. Hurtado. 2000b. Length-scale effects in nano- and micro-mechanics of solids. *Key Engineering Materials* 183–187: 1–7.

Kim, S. H., D. B. Asay, and M. T. Dugger. 2007. Nanotribology and MEMS. *NanoToday* 2: 22–29.

Kopycinska-Müller, M., R, H. Geiss, and D. C. Hurley. 2006. Contact mechanics and tip shape in AFM-based nanomechanical measurements. *Ultramicroscopy* 106: 466–474.

Krim, J. 2002. Surface science and the atomic scale origins of friction: What once was old is new again. *Surface Science* 500: 741-758.

Lantz, M. A., S. J. O'Shea, A. C. F. Hoole, and M. E. Welland. 1997a. Lateral stiffness of the tip and tip-sample contact in frictional force microscopy. *Appl. Phys. Lett.* 70: 970–972.

Lantz, M. A., S. J. O'Shea, M. E. Welland, and K. L. Johnson. 1997b. Atomic-force-microscope study of contact area and friction on $NbSe_2$. *Phys. Rev. B* 55: 10776–10784.

Lantz, M. A., D. Wiesmann, and B. Gotsmann. 2009. Dynamic superlubricity and the elimination of wear on the nanoscale. *Nature Nanotech.* 4 586–589.

Li, Q., K.-S. Kim, and A. Rydberg. 2006. Lateral force calibration of an atomic force microscope with a diamagnetic levitation spring system. *Rev. Sci. Instrum.* 77: 065105.
See also the online resource http://www.engin.brown.edu/facilities/nanomicro/calibrator/dlfcindex.html for instructions on implementing the technique.

Lin, S.C., and K. Banerjee. 2008. Cool chips: Opportunities and implications for power and thermal management. *IEEE Trans. Electron Devices* 55: 245–255.

Luan, B., and M. O. Robbins. 2005. The breakdown of continuum models for mechanical contacts. *Nature* 435: 929–932.

Luan, B., and M. O. Robbins. 2006. Contact of single asperities with varying adhesion: Comparing continuum mechanics to atomistic simulations. *Phys. Rev. E* 74: 026111.

Luo, J. K., Y. Q. Fu, H. R. Le, J. A. Williams, S. M. Spearing, and W. I. Milne. 2007. Diamond and diamond-like carbon MEMS. *J. Micromech. Microeng.* 17: S147–S163.

Maboudian R., and C. Carraro. 2003. Surface engineering for reliable operation of MEMS devices. *J. Adh. Sci. Technol.* 17: 583–591.

Maboudian, R., and R. T. Howe. 1997. Critical review: Adhesion in surface micromechanical structures. *J. Vac. Sci. Technol. B* 15: 1–20.

Major, R. C., H. I. Kim, J. E. Houston, and X.-Y. Zhu. 2003. Tribological properties of alkoxyl monolayers on oxide terminated silicon. *Tribol. Lett.* 14: 237–244.

Major, R. C., J. E. Houston, M. J. McGrath, J. I. Siepmann, and X.-Y. Zhu. 2006. Viscous water meniscus under nanoconfinement. *Phys. Rev. Lett.* 96: 177803.

Marti, O., J. Colchero, and J. Mlynek. 1990. Combined scanning force and friction microscopy of mica. *Nanotechnology* 1: 141–144.

Mate, C. M., G. M. McClelland, R. Erlandsson, and S. Chiang. 1987. Atomic-scale friction of a tungsten tip on a graphite surface. *Phys. Rev. Lett.* 59: 1942–1945.

Maugis, D. 1992. Adhesion of spheres: The JKR-DMT transition using a Dugdale model. *J. Colloid Interface Sci.* 150: 243–269.

Meyer, E., R. Overney, K. Dransfeld, and T. Gyalog. 1998. *Nanoscience: Friction and Rheology on the Nanometer Scale.* Singapore: World Scientific.

Meyer, G., and N. M. Amer. 1988. Novel optical approach to atomic force microscopy. *Appl. Phys. Lett.* 53: 1045–1047.

Meyer, G., and N. M. Amer. 1990a. Optical-beam-deflection atomic force microscopy: The NaCl (001) surface. *Appl. Phys. Lett.* 56: 2100–2101.

Meyer, G., and N. M. Amer. 1990b. Simultaneous measurement of normal and lateral forces with an optical-beam-deflection atomic force microscope. *Appl. Phys. Lett.* 57: 2089–2091.

Mo, Y., K. T. Turner, and I. Szlufarska. 2009. Friction laws at the nanoscale. *Nature* 457: 1116–1119.

Morita, S., S. Fujisawa, and Y. Sugawara. 1996. Spatially quantized friction with lattice periodicity. *Surface Science Reports* 23: 1–41.

Munz, M. 2010. Force calibration in lateral force microscopy: A review of the experimental methods. *J. Phys. D: Appl. Phys.* 43: 063001.

Müser, M. H., and M. O. Robbins. 2000. Conditions for static friction between flat crystalline surfaces. *Phys. Rev. B* 61: 2335–2342.

Müser, M. H., L. Wenning, and M. O. Robbins. 2001. Simple microscopic theory of Amontons's laws for static friction. *Phys. Rev. Lett.* 86: 1295–1298.

Obraztsov, A. N., P. G. Kopylov, B. A. Logynov, M. A. Dolganov, R. R. Ismagilov, and N. V. Savenko. 2010. Single crystal diamond tips for scanning probe microscopy. *Rev. Sci. Instrum.* 81: 013703.

Ogletree, D. F., R. W. Carpick, and M. Salmeron. 1996. Calibration of frictional forces in atomic force microscopy. *Rev. Sci. Instrum.* 67: 3298–3306.

Olivero, P., S. Rubanov, P. Reichart, B. C. Gibson, S. T. Huntington, J. R. Rabeau, A. D. Greentree, J. Salzman, D. Moore, D. N. Jamieson, and S. Prawer. 2006. Characterization of three-dimensional microstructures in single-crystal diamond. *Diamond Rel. Mater.* 15: 1614–1621.

Park, J. Y, D. F. Ogletree, P. A. Thiel, and M. Salmeron. 2006. Electronic control of friction in silicon pn junctions. *Science* 313: 186.

Park, J. Y., Y. Qi, D. F. Ogletree, P. A. Thiel, and M. Salmeron. 2007. Influence of carrier density on the friction properties of silicon pn junctions. *Phys. Rev. B* 76: 064108.

Peiner, E., A. Tibrewala, R. Bandorf, H. Lüthje, L. Doering, and W. Limmer. 2007. Diamond-like carbon for MEMS. *J. Micromech. Microeng.* 17: S83–S90.

Persson, B. N. J. 2000. *Sliding Friction: Physical Principles and Applications.* Berlin: Springer-Verlag.

Pethica, J. B., and D. Tabor. 1979. Contact of characterized metal surfaces at very low loads: Deformation and adhesion. *Surf. Sci.* 89: 182–190.

Phinney, L. M., K. Fushinobu, and C. L. Tien. 2000. Subpicosecond laser processing of polycrystalline silicon microstructures. *Microscale Thermophys. Eng.* 4: 61–75.

Piétrement, O., J. L. Beaudoin, and M. Troyon. 1999. A new calibration method of the lateral contact stiffness and lateral force using modulated lateral force microscopy. *Tribol. Lett.* 7: 213–220.

Piétrement, O., and M. Troyon. 2001a. Study of the interfacial shear strength pressure dependence by modulated lateral force microscopy. *Langmuir* 17: 6540–6546.

Piétrement, O., and M. Troyon. 2001b. Quantitative study of shear modulus and interfacial shear strength by combining modulated lateral force and magnetic force modulation microscopies. *Surf. Interf. Anal.* 31: 1060–1067.

Reinstädtler, M., T. Kasai, U. Rabe, B. Bhushan, and W. Arnold. 2005. Imaging and measurement of elasticity and friction using the TR mode. *J. Phys. D: Appl. Phys.* 38: R269–R282.

Reitsma, M. G. 2007. Lateral force microscope calibration using a modified atomic force microscope cantilever. *Rev. Sci. Instrum.* 78: 106102.

Rogers, J. W., and L. M. Phinney. 2002. Nanosecond laser repair of adhered MEMS structures. *J. Heat Transfer* 124: 394–396.

Romig, A. D., Jr., M. Y. Dugger, and P. J. McWhorter. 2003. Materials issues in microelectromechanical devices: Science, engineering, manufacturability and reliability. *Acta Materialia* 51: 5837–5836.

Russick, E. M., C. L. J. Adkins, and C. W. Dyck. 1997. Supercritical carbon dioxide extraction of solvent from micromachined structures. *Supercritical Fluids,* Ch. 18, pp. 255-269. M. A. Abraham and A. K. Sunol, Eds. American Chemical Society, Washington D.C.

Sader, J. E. 1999. Calibration of rectangular atomic force microscope cantilevers. *Rev. Sci. Instrum.* 70: 3967–3969.
    See also http://www.ampc.ms.unimelb.edu.au/afm/calibration.html for an online resource for calculating normal and lateral spring constants of rectangular cantilevers.

Sader, J. E., and C. P. Green. 2004. In-plane deformation of cantilever plates with applications to lateral force microscopy. *Rev. Sci. Instrum.* 75: 878–883.

Sader, J. E., J. Pacifico, C. P. Green, and P. Mulvaney. 2005. General scaling law for stiffness measurement of small bodies with applications to the atomic force microscope. *J. Appl. Phys.* 97: 124903.

Schwarz, U. D. 2003. A generalized analytical model for the elastic deformation of an adhesive contact between a sphere and a flat surface. *J. Coll. Interf. Sci.* 261: 99–106.

Schwarz, U. D., O. Zwörner, P. Köster, and R. Wiesendanger. 1997. Preparation of probe tips with well-defined spherical apexes for quantitative scanning force microscopy. *J. Vac. Sci. Technol. B* 15: 1527–1530.

Sekaric, L., J. M. Parpia, H. G. Craighead, T. Feygelson, B. H. Houston, and J. E. Butler. 2002. Nanomechanical resonant structures in nanocrystalline diamond. *Appl. Phys. Lett.* 81: 4455–4457.

Shan, Z. W., R. K. Mishra, S. A. S. Asif, O. L. Warren, and A. M. Minor. 2008. Mechanical annealing and source-limited deformation in submicrometre-diameter Ni crystals. *Nature* 7: 115–119.

Socoliuc, A., E. Gnecco, S. Maier, O. Pfeiffer, A. Baratoff, R. Bennewitz, and E. Meyer. 2006. Atomic-scale control of friction by actuation of nanometer-sized contacts. *Science* 313: 207–210.

Stachowiak, G. W., and A. W. Batchelor. 2005. *Engineering Tribology* (3rd ed.). Boston: Elsevier Butterworth-Heinemann.

Stiernstedt, J., M. W. Rutland, and P. Attard. 2005. A novel technique for the in situ calibration and measurement of friction with the atomic force microscope. *Rev. Sci. Instrum.* 76: 262508.

Sullivan, J. P., T. A. Friedmann, and K. Hjort. 2001. Diamond and amorphous carbon MEMS. *MRS Bulletin* 26: 309–311.

Szlufarska, I., M. Chandross, and R. W. Carpick. 2008. Recent advances in single-asperity nanotribology. *J. Phys. D: Appl. Phys.* 41: 123001.

Tabor, D. 1977. Surface forces and surface interactions. *J. Coll. Interf. Sci.* 58: 2–13.

Takai, Y., T. Kawasaki, Y. Kimura, T. Ikuta, and R. Shimizu. 2001. Dynamic observation of an atom-sized gold wire by phase electron microscopy. *Phys. Rev. Lett.* 87: 106105.

Tocha, E., H. Schönherr, and G. J. Vancso. 2006. Quantitative nanotribology by AFM: A novel universal calibration platform. *Langmuir* 22: 2340–2350.

Tortonese, M., and M. Kirk. 1997. Characterization of application specific probes for SPM. *SPIE* 3009: 53–60.

Unertl, W. N. 1999. Implications of contact mechanics models for mechanical properties measurements using scanning force microscopy. *J. Vac. Sci. Technol. A* 17: 1779–1786.

Varenberg, M., I. Etsion, and G. Halperin. 2003. An improved wedge calibration method for lateral force in atomic force microscopy. *Rev. Sci. Instrum.* 74: 3362–3367.

Villarrubia, J. S. 1996. Scanned probe microscope tip characterization without calibrated tip characterizers. *J. Vac. Sci. Technol. B* 14: 1518–1521.

Villarrubia, J. S. 1997. Algorithms for scanned probe microscope image simulation, surface reconstruction, and tip estimation. *J. Res. Natl. Inst. Stand. Technol.* 102: 425–454.

Wahl, K. J., S. V. Stepnowski, and W. N. Unertl. 1998. Viscoelastic effects in nanometer-scale contacts under shear. *Tribol. Lett.* 5: 103–107.

Watson, G. S., B. P. Dinte, J. A. Blach-Watson, and S. Myhra. 2004. Friction measurements using force versus distance friction loops in force microscopy. *Appl. Surf. Sci.* 235: 38–42.

Williams, J. A. 1994. *Engineering Tribology.* Oxford, UK: Oxford University Press.

Williams, J. A. 2006. Tribology and MEMS. *J. Phys. D: Appl. Phys.* 39: R201–R214.

6 Effects of Micro- and Nanoscale Texturing on Surface Adhesion and Friction

Min Zou

CONTENTS

6.1 INTRODUCTION

Surface texture is the repetitive or random deviation from the nominal surface that forms the 3-D topography of the surface. Surface texture includes mainly roughness and waviness. Roughness, or micro- and nanoroughness, is formed by fluctuations in the surface of short wavelengths and is characterized by hills (local maximum or asperities) and valleys (local minimum) of varying amplitude and spacing. Waviness, or macro-roughness, is the surface irregularity of longer wavelengths.

Surface-texturing or surface roughness modification has long been recognized as an effective approach to reducing the contact area between surfaces and therefore reducing the adhesion and friction forces between interfaces. Cylinder liner honing is one of the earliest and most familiar macroscale commercial applications of surface texturing (Willis 1985). Recently, laser surface-texturing was used to reduce friction in mechanical seals, piston rings, and thrust bearings (Etsion 2005). As mechanical systems shrink in size, adhesion and friction become the main issues affecting the reliability of miniaturized systems due to the increased surface forces caused by dramatically increased surface-to-volume ratio (Komvopoulos 2003; Maboudian and Howe 1997). Therefore, surface-texturing is of critical importance for overcoming the adhesion and stiction in miniaturized systems such as micro-electromechanical systems (MEMS) (Komvopoulos 2003; Maboudian and Howe 1997; Zhu et al. 2006) and in magnetic storage devices (Raeymaekers, Etsion, and Talke 2007; Chilamakuri and Bhushan 1997; Liu 1997). Adhesion and friction forces in these systems are affected mainly by surface forces, such as capillary, electrostatic, and van der Waals forces (Komvopoulos 2003), which are largely dependent on the contact geometry and surface topography. Laser surface-texturing with isolated microsized islands of 5–10 μm diameter has been successfully used in computer hard drives to improve tribological performance (Chilamakuri and Bhushan 1997; Liu 1997). Different etching processes have been developed to produce surface-texturing to improve tribological properties of MEMS (Alley et al. 1992; Yee, Chun, and Jong 1995; Houston, Maboudian, and Howe 1995). Surfaces produced by these etching methods have fractal geometry with random multiscale roughness, which may limit their benefits in MEMS/NEMS (Maboudian and Howe 1997).

Compared to the microroughness produced by laser and etching methods, nano-surface-texturing with isolated nanoislands has several unique potential advantages for tribological applications that involve ultrasmooth surfaces. First, the contact areas of nanotextured surfaces (NTSs) are smaller than that of microtextured surfaces (MTSs) due to the significantly reduced size of the asperities. This reduced contact area can result in more significant reduction of adhesion and friction forces. Second, the nanometer-sized asperities are generally much harder than their micron-sized counterparts because of their nearly defect-free nature (Gerberich et al. 2003), which can considerably enhance the wear resistance of a surface. Third, the presence of nano-sized asperities increases the surface hydrophobicity of hydrophobic surfaces (Choi, Kim, and Kim 2004), which will lead to the reduction of meniscus-mediated adhesion and friction forces in a humid environment. Therefore, NTSs have greater potential to reduce catastrophic failures in MEMS/NEMS devices involving contacting surfaces.

Due to their anticipated tribological advantages, NTSs have attracted much attention recently (Song et al. 2010; Nair and Zou 2008; Nair et al. 2008; Wang et al. 2007; Zou et al. 2006; Zou, Cai, and Wang 2006; Zou et al. 2005). Studies that compare the adhesion and friction performances of a NTS and MTSs provided experimental support for the anticipated benefits of NTSs (Zou, Seale, and Wang 2005).

This chapter reviews our recent work on micro- and nanoscale surface-texturing for potential tribological applications in miniaturized systems. In the following sections, fabrication and adhesion and friction studies of various micro- and nanoscale textured surfaces (MNTSs) will be described in detail. MNTSs covered with self-assembled monolayers are also included because the combination of surface-texturing and chemical modification can alter the surface wetting properties and thus change the tribological properties.

6.2 EXPERIMENTAL TECHNIQUES

6.2.1 FABRICATION OF MICRO- AND NANOTEXTURED SURFACES

Many techniques can be used to produce large-scale microtextures on surfaces, such as laser surface-texturing (Etsion 2005) and photolithography (Singh et al. 2009), and so on. Generating large-scale nanotextures on surfaces, however, is very challenging and could be very costly. Various techniques were investigated for producing NTSs. Nanolithography techniques, such as electrobeam lithography (Di Fabrizio et al. 2003), interference lithography (Murillo et al. 2005), nanoimprint lithography (Kono et al. 2005), and soft lithography (Yao et al. 2004), can be used to create nano-sized topographical and material variations on substrate surfaces with specific patterns. A rich variety of other methods, such as physical vapor deposition (Singh, Singh, and Srivastava 2003), chemical vapor deposition (Huh et al. 2005), dip coating (Tae-Sik Yoon et al. 2004), and spin coating (Fu-Ken Liu et al. 2003), can be used to produce NTSs with randomly distributed nanostructures. Each of these techniques has its own advantages and disadvantages and may be best suited for specific applications. Many of these techniques may not be suitable for MEMS/NEMS applications due to MEMS/NEMS process constraints. To address this issue, four types of techniques of fabricating MNTSs for potential MEMS/NEMS applications were developed and will be described in detail in this review. They are rapid aluminum-induced crystallization (AIC) of amorphous silicon (a-Si), UV-assisted crystallization of a-Si, spin coating of colloidal silica nanoparticle solution, and the anodized aluminum oxide (AAO) template method. The first three techniques can be used to fabricate MTSs and NTSs with random textures and the AAO template method can be used to produce nanodot-patterned surfaces (NDPSs) with ordered nanodots. It is noted here that randomly textured surfaces are desirable for tribological applications due to lower propensity of causing resonance issues, while patterned surfaces are ideal for fundamental studies of the tribological and mechanical properties of textured surfaces. These four micro- and nano-surface-texturing methods are described below in detail.

6.2.1.1 Micro- and Nanotextured Surfaces by Rapid Aluminum-Induced Crystallization of Amorphous Silicon

AIC of a-Si has been studied extensively during the past two decades for various applications such as thin-film transistors, sensors, solar cells, and display panels (Nast and Hartmann 2000; Gall et al. 2002; Klein et al. 2004). During a traditional AIC of a-Si process, a layer of Al is deposited on a substrate followed by deposition of a layer of a-Si of similar thickness using thin film deposition techniques. The resulting film structure is then subject to thermal annealing for an extended period, normally hours, to crystallize the a-Si. During the crystallization process, the Al beneath the a-Si diffuses to the top of the silicon layer. After cooling to ambient temperatures, the Al is then etched off by selective wet chemical etching, exposing the crystallized silicon surface. It is important to point out that traditional AIC of a-Si strives to grow large and continuous Si grains and is targeted for applications in electronics and optoelectronics.

We have developed a rapid AIC of a-Si process that can be used to produce surfaces textured by isolated micro- or nanoscale Si grains with controllable crystallite size, height, and density for potential tribological applications (Song et al. 2010; Nair and Zou 2008). The process steps involved in fabricating Si MNTSs are essentially the same as those used in traditional AIC of a-Si studies. However, there are several major differences in the process parameters and procedures. First, the Al is deposited on top of a-Si instead of underneath it. Second, the film thickness ratio of Al to a-Si is normally kept at about 1:1 in a traditional AIC of a-Si study, while this ratio needs to be much higher, from 4:1–8:1, in order to produce MNTSs. Third, the annealing temperature used in a traditional AIC of a-Si study is normally below the Al-Si eutectic temperature of 577°C with annealing duration longer than 0.5 h, while the annealing temperature needs to be above 577°C with an annealing time shorter than 20 s for fabricating MNTSs. Finally, the samples are normally placed in a furnace and heated during temperature ramp-up in a traditional AIC of a-Si study, while in our study the samples need to be placed in a furnace that has already reached the targeted annealing temperature for generating MNTSs. One unique advantage of this technique is that the spatial density and the heights of micro- and nanoislands can be adjusted by process parameters, such as a-Si and aluminum thicknesses, their deposition parameters, and annealing conditions.

A wide range of surface topographies can be produced by the rapid AIC of a-Si technique through varying the parameters involved in this process. The various steps of the AIC of a-Si process are shown in Figure 6.1. Single-crystal silicon wafers with a silicon oxide layer were used as substrates. The silicon oxide layer, thermally grown in a furnace at a temperature of 1100°C for 10 h, was used to prevent the crystalline orientation of the substrates from affecting the crystallization of a-Si process. An a-Si layer was then deposited on these substrates through plasma-enhanced

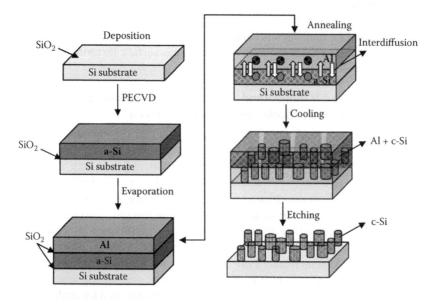

FIGURE 6.1   Process flow chart of surface texturing by rapid AIC of a-Si.

chemical vapor deposition (PECVD). The PECVD deposition parameters such as chamber pressure, radio frequency (RF) power, substrate temperature, and SiH$_4$ flow rates were kept constant at 1 Torr, 20 W, 250°C, and 85 sccm, respectively. The thickness of the a-Si was varied from 100 nm to 400 nm by controlling the PECVD deposition time. The samples were removed from the PECVD system after the a-Si was deposited and left in ambient air for three days to form a native silicon oxide layer on top of the a-Si film. Since the thickness of the native silicon oxide layer will barely change after three days, leaving the sample in air for three days ensured that the surfaces had consistent native silicon oxide layer thickness from one experiment to another. After allowing a consistent thin layer of native oxide to form, the samples were transferred to an evaporator, where an 800-nm-thick aluminum layer was deposited while keeping the pre-evaporation chamber pressure at 7.0×10^{-6} mbar and evaporation rate at 3 ~ 4 nm/sec. The samples were then annealed in a conventional furnace at different temperatures ranging from 600°C to 850°C with durations varying from 5 seconds to 20 seconds. The samples were taken out of the furnace immediately and cooled in air. The aluminum was then etched away by wet selective etching using the Aluminum Etchant – Type D solution from Transene Company, Inc. (Danvers, MA). During etching, the samples were immersed in the solution for about 15 minutes while maintaining the solution at 50°C. After aluminum etching, silicon nanostructures on the substrate were exposed.

6.2.1.2   Selectively Micro- and Nanotextured Surfaces by UV-Assisted Crystallization of Amorphous Silicon

Amorphous Si films of 100-nm thickness were deposited using PECVD on silicon wafers with thermally grown silicon oxide layers. The RF power, chamber pressure, substrate temperature, and SiH$_4$ flow rates were controlled at 15W, 0.5 Torr, 250°C, and 85 sccm, respectively. After a-Si deposition, 50-nm-thick Al was deposited on top of the a-Si using a thermal evaporator. The sample was then subjected to UV (λ = 365 nm) irradiation at the intensity of 10 mW/cm^2 for 10 seconds through a patterned photo mask. The UV light irradiated onto the samples was produced by filtered high-pressure mercury lamps. The samples were then immediately annealed in a N$_2$ environment at 400°C for 10 seconds. Finally, the remaining aluminum was removed using selective wet chemical etching as described in Section 6.2.1.1. After etching, micro- and nanotextures were formed on the surfaces in patterned areas (Zou et al. 2005).

6.2.1.3   Micro- and Nanotextured Surfaces by Spin Coating of Colloidal Silica Nanoparticle Solution

Spin coating has been widely used to deposit uniform photoresist films onto wafers in the semiconductor industry. However, this technique has not been studied by others for nano- and microtexturing of surfaces for tribological applications. In comparison with other nano- and micro-surface-texturing techniques, spin coating has several unique advantages for controlling mechanical and tribological properties of textured surfaces, including adjustable texture material, size, and density. Because various types of nanoparticles are commercially available, spin coating provides unique opportunities for fundamental research on the mechanical and tribological properties of various nanoparticles and their textured surfaces.

The spin coating of colloidal silica nanoparticle solution process was developed for producing MTSs and NTSs for tribological applications (Zou, Seale, and Wang 2005). Microtextured samples were prepared by spin coating diluted colloidal silica nanoparticle solution (SNOWTEX® from Nissan Chemical Industries) on Si substrates. The nanoparticle solutions were diluted with deionized (DI) water at mass ratios of 1:100 and 1:400, sonicated for 20 min, and spin coated onto silicon substrates at a spin speed of 6000 rpm/min for 1 min to create two different MTSs. The microtextures were formed by aggregations of the nanoparticles produced during spin coating. The samples were then annealed at 600°C in N_2 for 1 h to improve the adhesion between the nanoparticles and the substrates and slowly cooled to room temperature at a rate of about 3°C/min.

A nanotextured sample was produced by spin coating 1:160 diluted and sonicated nanoparticle solution onto a Si substrate at a spin speed of 4000 rpm/min for 1 min. To avoid large aggregations of the nanoparticles, DI water was dropped on the sample surface to wash away the nanoparticle solution 10 s after spinning started. The sample was then annealed at 900°C in N_2 for 2 h and slowly cooled to room temperature at a rate of about 3°C/min. The high-temperature anneal allowed the aggregated nanoparticles to melt into larger nanoscale particles and enhanced the adhesion between the nanoparticles and the substrate.

6.2.1.4 Nanodot-Patterned Surfaces by Anodized Aluminum Oxide Template Method

NDPSs are surfaces covered with orderly distributed arrays of nanodots with approximately the same size and shape. Fabrication of such nanodot arrays has attracted considerable attention recently because these nanostructures show novel physical properties that lead to potential unique applications in electronic, optoelectronics, and magnetic storage (Loo et al. 2002; Donthu et al. 2005; Jia-Yang Juang and Bogy 2005). Many techniques, such as electronbeam, interference, imprint, and soft lithographies, as well as self-assembled anodized aluminum oxide (AAO) template methods, have been used to fabricate NDPS (Di Fabrizio et al. 2003; Murillo et al. 2005; Kono et al. 2005; Yao et al. 2004; Choi et al. 2004; Chik et al. 2004). Among the existing techniques, the self-assembled AAO template technique has the advantages of being parallel and scalable; moreover, it can be combined with conventional deposition and/or etching processes to pattern surfaces with various nanostructured topologies and materials. For example, using the AAO templates as masks, various metal nanodot arrays with nearly identical dot size, height, and spacing can be fabricated by thermal evaporation of metal sources (Masuda and Satoh 1996; Masuda, Yasui, and Nishio 2000; Liang et al. 2002). The size and spacing of the nanodots can be controlled by varying the anodization conditions of the AAO templates, while the height of the nanodots can be partially controlled by the evaporation process parameters. These AAO template techniques are therefore particularly suitable for producing relatively large-area, nanopatterned surfaces for tribological applications in miniaturized systems.

Sample processing starts with an AAO template fabricated using a two-step anodization process on a pure (99.999%) Al foil described in detail elsewhere (Masuda and Fukuda 1995; Li et al. 1998). First, the Al foil was anodized in 0.3 M oxalic acid at 40 V and 1°C for 15 hours to grow a thick porous oxide layer. The resulting AAO film was then chemically stripped from the Al foil and a secondary anodization in 0.3 M oxalic acid at 40 V and 1°C for 5 min was carried out. Using this two-step technique, good ordering was obtained over micron-sized regions and resulted in an AAO film approximately 300 nm in thickness with 50-nm pore diameters spaced 100 nm apart. A through-hole mask was prepared by separating the AAO film from the Al foil in a saturated $HgCl_2$ solution and removing the bottom alumina barrier layer in 5 wt.% phosphoric acid at 30°C for 34 min. The remaining AAO layer was then lifted off onto a Si substrate. This resulted in a through-hole AAO mask approximately 3 × 3 mm in size weakly bonded by van der Waals force to a Si substrate.

Next, a 100-nm layer of Ni was deposited onto the Si through the AAO mask by thermal evaporation using an Edwards E306A system with base vacuum of ~10^{-6} Torr. The Ni source was placed directly underneath the sample with a sample-to-source distance of approximately 18 cm. This allowed a significant fraction of the evaporated Ni (with an estimated angular dispersion of ~100 mm/5 mm = 20:1) to travel down the pores of the mask (which had an aspect ratio of 300 nm/50 nm = 6:1). Under these conditions, a 120-nm-thick layer of Ni coated on top of the AAO mask results in 75-nm-high dots. After evaporation, the AAO film was removed by means of a wet chemical etch (a 1:1 mixture of 6 wt.% phosphoric acid: 1.8 wt.% chromic acid), leaving behind a well-ordered array of conical-shaped Ni dots (Zou et al. 2006).

6.2.2 ADHESION AND FRICTION STUDIES OF MICRO- AND NANOTEXTURED SURFACES

Adhesion and friction investigations were conducted in air at a relative humidity of about 40% using the TriboIndenter (Hysitron, Inc.), which has force and displacement sensing capabilities in both vertical and lateral directions. Figure 6.2 shows the schematics of the sensing system of the TriboIndenter, which consists of two three-plate capacitive force–displacement transducers with integrated electrostatic actuation functionality. The forces are applied electrostatically, and the displacements are measured with a differential capacitor technique. The resolution is 0.02 nm with a 0.2-nm noise floor for the vertical displacement, 3 nN with a 100-nN noise floor for the vertical/indentation force, and 500 nN with a 5-μN noise floor for the lateral force, respectively. The TriboIndenter also has an optical microscope and a scanning probe microscope (SPM) component, allowing in-situ imaging of a sample surface and accurately locating specific areas of interest, such as textured areas and individual micro- and nanoislands. Conical diamond tips of different radii of curvature attached to the transducer were used to measure the sample surface topography, adhesion force, and friction force.

Adhesion tests were performed using the TriboIndenter by controlling a diamond tip to follow a predefined displacement profile (Zou et al. 2005). A typical

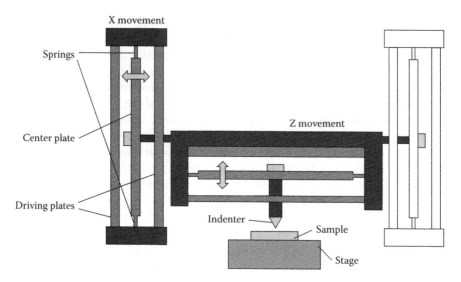

FIGURE 6.2 Schematics of the TriboIndenter sensing system. From M. Zou, L. Cai, and H. Wang. Adhesion and friction studies of a nano-textured surface produced by spin coating of colloidal silica nanoparticle solution. *Tribology Letters* 21, no. 1 (2006): 25–30, Figure 1. With kind permission from Springer Science+Business Media.

displacement profile includes a tip approaching a sample surface from 50 nm above the sample, indenting 1–80 nm into the sample, and then withdrawing from the sample. To illustrate the testing process, a conical tip with 100-μm radius of curvature was used to indent a textured surface to 1-nm indentation depth. Figure 6.3 shows a force–displacement curve during an adhesion measurement as the tip was brought into contact with the sample and then was withdrawn from the sample.

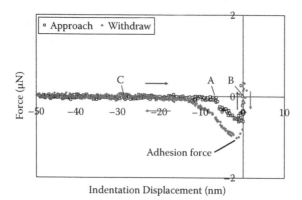

FIGURE 6.3 Force-displacement curve obtained during an adhesion measurement. From M. Zou, L. Cai, H. Wang, D. Yang, and T. Wyrobek. Adhesion and friction studies of a selectively micro/nano-textured surface produced by UV assisted crystallization of amorphous silicon. *Tribology Letters* 20 (2005): 43–52, Figure 2. With kind permission from Springer Science+Business Media.

While approaching, the tip first experiences an attractive normal force at point A. As the tip further approaches the sample, the attractive normal force increases first, then decreases, eventually turns repulsive, and at point B, goes through zero force. As the tip is withdrawn, after breaking the solid–solid adhesion (minimum force in the curve), the attractive force gradually decreases to zero at point C. The adhesion force is the minimum force while the tip is withdrawing.

The friction performance of these surfaces was tested under various normal loads at a sliding speed of 1 µm/s using diamond tips with 5- and 100-µm radii of curvature (Zou et al. 2005). After engaging a tip with the sample surface at a contact force of 2 µN, the desired constant normal load was applied. The tip then slides across the sample surface at a desired speed under the applied load during each experiment. The normal and lateral displacements and the normal and lateral forces were recorded simultaneously and continuously as a function of time during sliding for each operation condition tested. The coefficient of friction (COF), defined as the ratio of the measured lateral force to applied normal force, was calculated.

Variable load friction/scratch tests were also conducted using a 1-µm diamond tip employing a ramp load profile from 0 µN to ten different maximum normal loads (10 µN to 500 µN) at a sliding speed of 1 µm/s with 8-µm scratch lengths (Wang et al. 2007). The friction/scratch tests consisted of seven steps: (1) the tip engaging the sample surface at a contact force of about 1 µN in the midpoint of the expected scratch, (2) the tip sliding toward one end of the expected scratch in 4 s under zero normal load, (3) the tip staying at this end of the scratch for 5 s under zero normal load, (4) the tip scratching toward the other end of the expected scratch in 8 s under the desired ramp normal load, (5) the tip staying at the end of the scratch for 5 s while the normal load is reduced to zero, (6) the tip moving back to the middle of the scratch under zero normal load, and (7) the tip withdrawing from the sample surface. The purpose of employing steps (1) and (2) was to properly account for the sample tilt in the normal displacement data collected. The time allowed in step (3) was to minimize any dynamic effects from step (2). The friction/scratch tests were repeated three times for each ramp load profile. The normal and lateral displacements and forces were recorded simultaneously and continuously as a function of time during sliding for each test. The COF was calculated as the ratio of the measured lateral force and applied normal force during tip sliding in step (4).

6.2.3 Wetting Property Study of Micro- and Nanotextured Surfaces

Surface wettability, or wetting property, is the ability of any solid surface to be wetted when in contact with a liquid. The wetting property can be characterized by the contact angle; that is, the angle between the liquid–solid interface and the liquid–vapor interface. Low contact angles reflect good wetting conditions, and high contact angles reflect nonwetting conditions.

The contact angle mostly depends on the surface tension of the test liquid and the surface energy of the solid substrate. Higher contact angles are generally obtained when the surface tension of the test liquid is higher and the surface energy of the solid substrate is lower. Typically, water is used to measure the contact angle of a surface. Water contact angle (WCA) analysis is a convenient way of quantifying the behavior

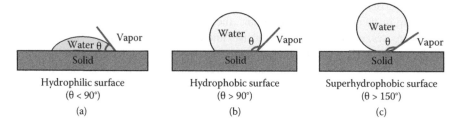

FIGURE 6.4   WCA, θ, of (a) a hydrophilic surface, (b) a hydrophobic surface, and (c) a superhydrohobic surface.

of water in contact with solid surfaces and the formation of droplets. As shown in Figure 6.4, by measuring the angle between the water–solid and the water–vapor interfaces, the wetting property of the surface can be determined. If the WCA is less than 90°, the surface is called *hydrophilic*, and the water tends to spread over the surface. If the WCA is greater than 90°, the surface is called *hydrophobic*, and the water tends to form droplets on the surface. If the WCA is greater than 150°, the surface is called *superhydrophobic*, and the water forms a near-spherical droplet on the surface.

A video-based contact angle measurement system (VCAMS) (OCA 15 plus, DataPhysics Instruments GmbH) was used to characterize the WCAs of the sample surfaces (Song et al. 2010). Static WCAs were measured using the sessile drop method dispensing 0.5- to 7-μl drops of DI water on the sample surfaces. Water sliding angles were measured by dropping a water droplet on a sample positioned on a tilting stage. The tilting angle of the stage was adjusted by a micrometer with a resolution of 0.03°. The measurements were repeated three to five times for each sample at different locations on the sample surfaces. All WCA measurements were taken under ambient laboratory conditions with about 20°C temperature and about 45% relative humidity.

6.3   ADHESION AND FRICTION PERFORMANCE OF MICRO- AND NANOTEXTURED SURFACES

6.3.1   ADHESION AND FRICTION PERFORMANCES OF MICRO- AND NANOTEXTURED SURFACES

In this section, the adhesion and friction performance of the MNTSs fabricated by the four different techniques described in Section 6.2.1 are presented and compared with the corresponding smooth control surfaces. Comparison will also be made among the MNTS surfaces fabricated by the four techniques as well as between a NTS and two MTSs.

6.3.1.1   Adhesion and Friction Performance of Nanotextured Surfaces by Rapid Aluminum-Induced Crystallization of Amorphous Silicon

Two samples fabricated by rapid AIC of a-Si were selected for tribological studies to understand the effect of surface nanotopography on the adhesion and friction performance of these surfaces (Nair and Zou 2008). Sample 1 was fabricated using 100-nm-thick a-Si, and sample 2 was fabricated using 400-nm-thick a-Si; both have 800-nm Al on top of the a-Si and were annealed at 700°C for 5 s. The topographies of the two samples were analyzed using both scanning electron microscopy (SEM) and atomic force microscopy (AFM). SEM micrographs in Figure 6.5 show that after the rapid AIC of a-Si processes, the sample surfaces were textured by irregularly shaped nanoscale silicon islands. Energy-dispersive x-ray spectroscopy (EDS) and x-ray diffraction (XRD) analysis confirmed these islands to be made of silicon. The average sizes of the textures, the texture area coverage, and the texture count are 0.028 μm^2 and 0.070 μm^2, 12.5% and 18.1%, and 1567 and 919, respectively, for samples 1 and 2 (analyzed using ImageJ from National Institute of Health). AFM measurements show that the average texture heights are 276 nm and 461 nm, respectively, for samples 1 and 2.

Adhesion tests were carried out using the Hysitron TriboIndenter. Results from the adhesion tests of the two NTSs and a smooth Si surface using the 100-μm tip, along with the adhesion tests on the smooth Si surface using a 5-μm tip, are presented in Figure 6.6(a). It can be seen that the NTSs have much smaller adhesion forces than the smooth Si surface when using the 100-μm tip for the tests. When the

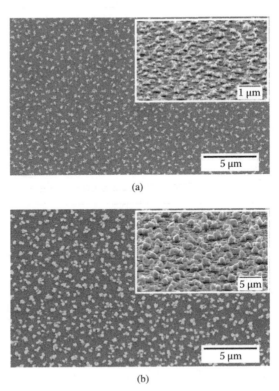

(a)

(b)

FIGURE 6.5 SEM micrographs of two NTSs fabricated by rapid AIC of a-Si selected for adhesion and friction studies: top views of (a) sample 1 and (b) sample 2. Insets are 75° oblique-angle views of the samples.

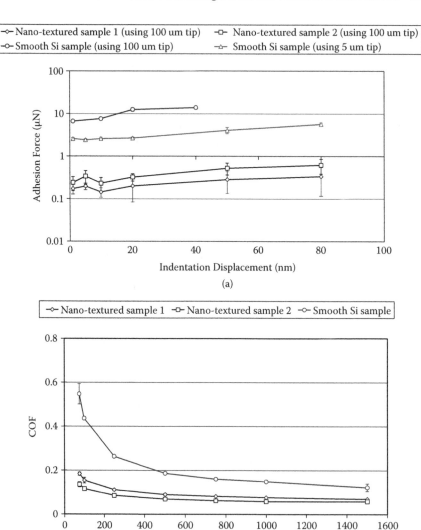

FIGURE 6.6 Results of (a) adhesion and (b) friction tests, showing reduced adhesion force and coefficient of friction on the NTS samples compared to those of a smooth Si surface. Data adapted from R. P. Nair and Min Zou, Surface-nano-texturing by aluminum-induced crystallization of amorphous silicon. *Surface & Coatings Technology* 203, no. 5–7 (2008): 675–679.

5-μm tip was used, the area of contact between the tip and the smooth Si surface was reduced in comparison to the 100-μm tip, thus decreasing the adhesion force; the latter is still significantly greater than that of NTSs tested by the 100-μm tip. Also, the adhesion forces of sample 1 were less than those of sample 2. This can be explained by the lower texture area coverage on the surface of sample 1. The adhesion performance difference should not depend on the texture height because the average texture heights were at least 276 nm, much larger than the maximum indentation displacement of 80 nm. Slightly increased adhesion forces and large variations in adhesion forces were also observed as indentation displacements increased. This can be understood from the SEM micrographs of the two NTS samples taken at a 75° oblique-angle shown in Figure 6.5 insets. It can be seen that samples have sharp textures of varying heights. As the indentation displacement increased, the sharp textures could be deformed under pressure, or they could cause the diamond tip to slip and then land on more textures at lower heights during an indentation, which resulted in increased adhesion forces and variations in adhesion forces.

Friction tests of the two selected samples and a sample cut from a bare silicon wafer (smooth Si) were performed using the TriboIndenter employing a 100-μm tip. Results from the friction tests are shown in Figure 6.6(b), where the relationships between the normal load and the COF for the above three samples were plotted. It can be seen that NTSs had much smaller COFs compared to the smooth silicon surface at all loads. The percentage improvement in the COF of samples 1 and 2 over the smooth Si sample were 66% and 75%, respectively, at 75 μN normal load and 43% and 52%, respectively, at 1500 μN normal load. The lower COFs of the NTSs were due to the smaller area of contact between the tip and the textured surfaces compared to the area of contact between the tip and the smooth surface. At the micro- and nanoscale, adhesion and plastic deformation both contribute to the friction force. As the normal load increases, the adhesion force increases sublinearly, thus causing the COF to decrease. Since the adhesion in the textured samples was low, it contributed less to the friction force. As a result, as the normal load increased, the relative improvement in the frictional performance of NTS samples over the smooth Si sample decreased.

Sample 1 had a higher COF than sample 2, even though it had less texture area coverage. This can be explained by the large deformation (approximately 250 nm) indicated by the measured normal displacements of the textures under the applied normal loads. Because the average height of the textures on sample 1 was 276 nm, close to the deformation of the textures, the tip could be interacting with the substrate between the textures, and thereby increase the surface area of contact and the intermolecular forces. However, the improvement in the frictional performance over the smooth surface was still 43%. Sample 2 had a greater improvement of 52% over the smooth Si sample because the average particle height of sample 2 (461 nm) was greater than that of sample 1. Therefore, texture height played an important role in the resulting COF of the NTSs.

6.3.1.2 Adhesion and Friction Performance of Selectively Micro- and Nanotextured Surfaces by UV-Assisted Crystallization of Amorphous Silicon

Selectively, MNTSs were formed on silicon substrates by UV-assisted crystallization of a-Si as described in Section 6.2.1.2 (Zou et al. 2005). Figure 6.7 shows an optical micrograph of a selectively textured surface. The darker areas, corresponding to UV-exposed regions, were textured by silicon crystallites formed during the AIC

FIGURE 6.7   Optical micrograph of a selectively micro- and nanotextured surface fabricated by UV-assisted crystallization of a-Si. The darker areas correspond to textured regions, while the brighter areas were not textured. Insets are SEM micrographs of the textured (top) and nontextured (bottom) surface areas.

of a-Si process, while the brighter areas, not exposed to UV, were the uncrystallized smooth a-Si surfaces. To visualize detailed surface structures of both textured and nontextured regions, SEM images were taken and are shown in the insets of Figure 6.7. The top inset illustrates that silicon crystallites are randomly distributed across UV-irradiated areas. The bottom inset shows that the area not irradiated by UV is very smooth. The surface roughnesses, measured by SPM with a conical tip of 1 μm nominal tip radius of curvature, are 27.7 nm and 0.4 nm for the textured and nontextured surfaces, respectively. Glancing incidence x-ray diffraction showed a peak around $2\theta = 28.5°$, corresponding to Si (111), indicating that crystallization occurred and the observed textured areas are indeed silicon crystallites.

Adhesion tests were performed on both textured and nontextured areas of the sample using a diamond tip of 100 μm radius of curvature. Figure 6.8(a) shows a comparison of average adhesion force of three measurements under progressively increased maximum indentation displacements for the textured surface and a smooth silicon (100) surface. We used silicon (100) for comparison because the adhesion forces for the nontextured a-Si surface were so large that they prevented the tip from lifting off from the nontextured surface, and therefore caused difficulties in making the correct adhesion force measurements. The adhesion force of the a-Si is expected to be larger than that of Si (100) because a-Si is not conductive and potentially has larger electrostatic charge buildup, which can also contribute to the adhesion force. It can be seen that the adhesion forces of the textured surface were only about 15% of those of the smooth silicon (100) surface. The adhesion forces for the textured surface did not increase much as the indentation displacement increased from 1 nm to 40 nm, while those of the silicon (100) surface increased significantly.

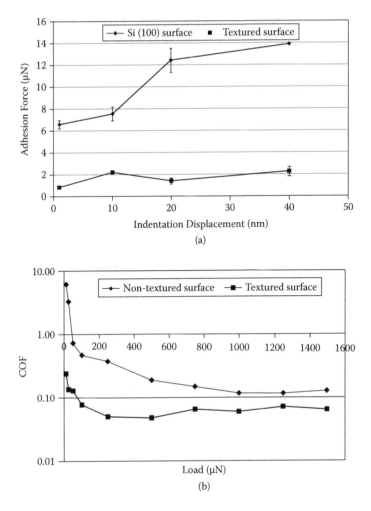

FIGURE 6.8 (a) Adhesion force comparison of textured surface area created by UV-assisted crystallization of a-Si and a smooth silicon surface. From M. Zou, L. Cai, H. Wang, D. Yang, and T. Wyrobek. Adhesion and friction studies of a selectively micro/nano-textured surface produced by UV assisted crystallization of amorphous silicon. *Tribology Letters* 20 (2005): 43–52, Figure 9. With kind permission from Springer Science+Business Media. (b) Coefficient of friction comparison of textured and nontextured surface areas.

Figure 6.8(b) shows the relationship between the average COF of three measurements and the normal load for both textured and nontextured areas of the surface. The COF axis is plotted in base-10 logarithm scale to better visualize the COFs differences on the two surface areas at both low and high loads. It can be seen that surface-texturing has very significant effects on COF at low loads. For example, COF on the textured surface measured at 10 μN load was only 4% of that of the nontextured surface area. The benefit of texturing was still very significant at high loads (> 500 μN) with 50% COF reduction.

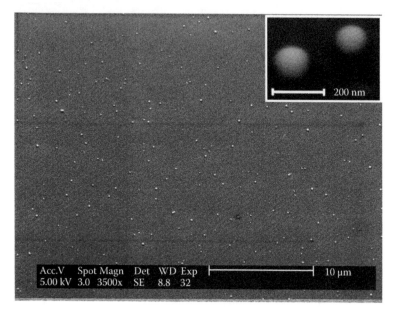

FIGURE 6.9 SEM images of the SNPT surface fabricated by spin coating of colloidal silica nanoparticle solution.

6.3.1.3 Adhesion and Friction Performance of a Nanotextured Surface by Spin Coating of Colloidal Silica Nanoparticle Solution

An NTS was fabricated on a silicon substrate by spin coating of colloidal silica nanoparticle solution as described in Section 6.2.1.3 (Zou, Cai, and Wang 2006). Figure 6.9 shows the SEM image of the silica nanoparticle–textured (SNPT) surface. The inset is an SEM image taken at a higher magnification showing details of the nanoparticles. It can be seen that the nanoparticles are of a spherical cap shape and are randomly distributed on the surface. Detailed analysis of the SEM images showed that the diameters of the nanoparticles are approximately between tens of nanometers to about 200 nm, and their heights are approximately 50% of their diameters. SPM measurements using a diamond tip of 40 nm nominal tip radius of curvature showed the average maximum peak heights of the SNPT and a silicon oxide film (SOF) surface to be 91.5 nm and 8.5 nm, respectively. The heights of individual nanotextures on the SNPT surface measured by the SPM are consistent with the SEM measurement results.

Adhesion tests were performed on the SNPT surface and the SOF surface using a diamond tip with 100 µm radius of curvature. Figure 6.10(a) shows a comparison of the average adhesion force of three measurements performed by a 100-µm tip with increasing maximum indentation displacements for the SNPT surface and a smooth silicon (100) surface. The reason for using a silicon (100) surface instead of the SOF surface for these measurements is that the adhesion forces for the SOF surface were so large that they prevented the tip from lifting off from the SOF surface and therefore caused difficulties in making correct adhesion force measurements. The adhesion force of the SOF surface is expected to be larger than that of a Si (100)

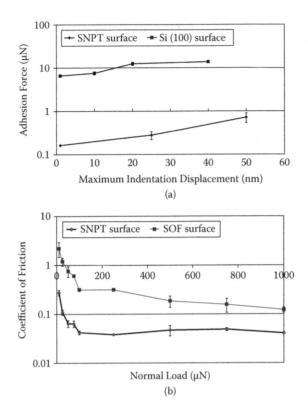

FIGURE 6.10   (a) Adhesion force comparison of a SNPT surface fabricated by spin coating of colloidal silica nanoparticle solution and a smooth silicon surface. (b) Coefficient of friction comparison of a SNPT surface and a SOF surface.

surface because silicon oxide is not conductive and potentially has a larger electrostatic charge buildup, which also contributes to the adhesion force. It can be seen that the adhesion forces of the SNPT surface were only 2–5% of those of the smooth silicon (100) surface.

Figure 6.10(b) shows the relationship between the average COF of three measurements and the normal load for both the SNPT and the SOF surfaces measured by a 100-μm tip. The COF axis is plotted in a base-10 logarithm scale for better visualization of the COF differences on the two surfaces at both low and high loads. It can be seen that nano-surface-texturing significantly affected the COFs. For example, at a low load (10 μN), the COF of the SNPT surface is only 12% of that of the SOF surface. At a high load (1000 μN), the COF of the SNPT surface is 33% of that of the SOF surface.

6.3.1.4   Adhesion and Friction Performance of a Nanodot-Patterned Surface by Anodized Aluminum Oxide Template Method

Ni NDPSs were fabricated by an AAO template in conjunction with thermal evaporation of Ni described in Section 6.2.1.4 (Zou et al. 2006). Both adhesion and friction

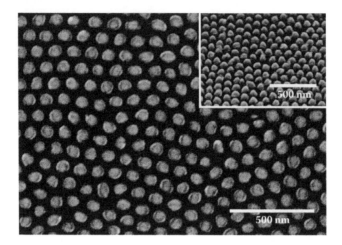

FIGURE 6.11 SEM micrographs of a Ni NDPS fabricated by AAO template method (top view). The inset is a 45° oblique-angle view. Hengyu Wang, R. Premachandran Nair, Min Zou, P. R. Larson, A. L. Pollack, K. L. Hobbs, M. B. Johnson, and O. K. Awitor. Friction study of a Ni nanodot-patterned surface. *Tribology Letters* 28 (2007): 183–189, Figure 1(a). With kind permission from Springer Science+Business Media.

properties of the Ni NDPS were studied. Ni was chosen because it is widely used in MEMS structures produced by LIGA (German acronym for lithography, electroplating, and molding), a micromachining technology used to produce MEMS made from metals, ceramics, or plastics (Hruby 2001). In addition, Ni NDPSs have promising potential applications in high-density magnetic recording media where tribological performance will also be a great concern.

Figure 6.11 shows a SEM micrograph of a Ni NDPS fabricated on a Si(100) substrate using the AAO template as a mask. The inset shows a 45° oblique angle view of the Ni nanodots. The hexagonal arrangement of the dots results from the hexagonal pore structure in the AAO template mask. The ordered Ni nanodots are approximately 75 nm in diameter with a dot-to-dot separation of 100 nm. The approximate dot height determined from this image was found to be 75 nm. From these images, it can be seen that the nanodots have conical shapes.

Adhesion tests were performed on a Ni NDPS and a smooth Si(100) surface using a diamond tip with a nominal radius of curvature of 100 μm. Figure 6.12(a) shows a comparison of the average adhesion forces of three measurements performed by the 100-μm tip under increasing maximum indentation displacements on the Ni NDPS and a Si(100) surface. It can be seen that nano-surface-patterning significantly reduced the adhesion forces of the Ni NDPS (up to a 92% reduction). For example, at a low indentation displacement (1 nm), the adhesion force of the Ni NDPS was only 16% of that of the Si(100) surface, and at a high indentation displacement (40 nm), the adhesion force of the Ni NDPS was only 8% of that of the Si(100) surface.

Figure 6.12(b) shows the relationship between the average COF of three measurements and the normal load for both the Ni NDPS and the Si(100) surface measured

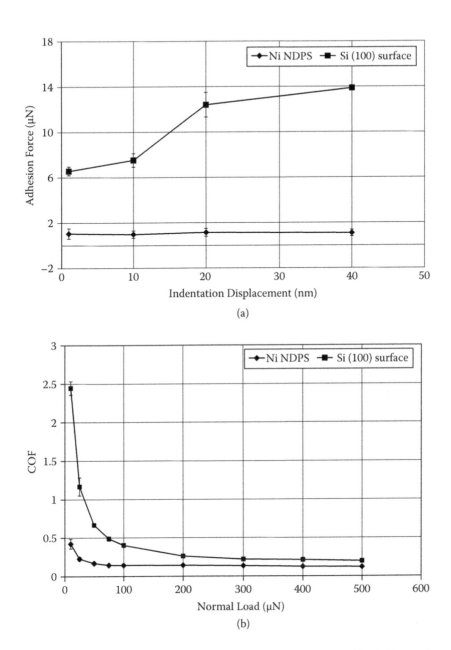

(a)

(b)

FIGURE 6.12 (a) Adhesion force comparison of a Ni NDPS fabricated by AAO template method and a smooth silicon surface. (b) Coefficient of friction comparison of the Ni NPS and the smooth Si surface. Hengyu Wang, R. Premachandran Nair, Min Zou, P. R. Larson, A. L. Pollack, K. L. Hobbs, M. B. Johnson, and O. K. Awitor. Friction study of a Ni nanodot-patterned surface. *Tribology Letters* 28 (2007): 183–189, Figures 4 and 5. With kind permission from Springer Science+Business Media.

by the 100-μm tip. It can be seen that nano-surface-patterning significantly reduced the COFs (up to an 83% reduction). For example, at a low load (10 μN), the COF of the Ni NDPS was only 17% of that of the Si(100) surface, and at a high load (500 μN), the COF of the Ni NDPS was 61% of that of the Si(100) surface.

6.3.2   COMPARISON OF ADHESION AND FRICTION PERFORMANCE OF VARIOUS MICRO- AND NANOTEXTURED SURFACES

The adhesion and friction performance of the NTSs and MTSs described in Section 6.3.1 (Nair and Zou 2008; Zou et al. 2006; Zou, Cai, and Wang 2006; Zou et al. 2005) were compared and plotted in Figure 6.13. It can be seen from Figure 6.13(a) that the adhesion forces of all textured surfaces are much smaller than that of smooth Si surface. The adhesion forces on all NTSs are lower than on MNTSs produced by UV-assisted crystallization of a-Si. NTSs produced by rapid AIC of a-Si and spin coating of colloidal silica nanoparticle are at a similar level. In contrast, adhesion forces of the Ni NDPSs produced by AAO template method are higher, most likely due to the softer Ni nanodots compared to silicon and silicon oxide nanotextures as well as larger surface area coverage of the Ni nanodots.

It can be seen from Figure 6.13(b) that the friction forces on smooth silicon, SOF, and a-Si surfaces are similar, and that they are much larger than those on all textured surfaces. The NTS produced by spin coating of colloidal silica nanoparticle has the lowest friction, followed by the MNTS produced by UV-assisted crystallization of a-Si, and then NTSs produced by AIC of a-Si. The low friction on the NTS produced by spin coating and the MNTS produced by UV-assisted crytallization of a-Si is likely due to the strong adhesion of those textures to the substrate. The Ni NDPS produced by the AAO template method showed the highest friction among the NTSs; again, most likely due to the softer Ni nanodots compared to silicon and silicon oxide nanotextures as well as larger surface area coverage of the Ni nanodots.

6.3.3   COMPARISON OF ADHESION AND FRICTION PERFORMANCE OF A SILICA NANOTEXTURED SURFACE WITH THOSE OF SILICA MICROTEXTURED SURFACES

6.3.3.1   Topographies of the Silica Nanotextured Surface and Silica Microtextured Surfaces

The adhesion and friction performance of an NTS and two MTSs were compared in order to provide experimental evidence of the anticipated benefits of NTSs (Zou, Seale, and Wang 2005). Both NTS and MTSs were produced by spin coating of colloidal silica nanoparticle solution on silicon substrates as described in Section 6.2.1.3.

SEM images of the silica NTS and MTSs show that all surfaces are covered with randomly distributed nanoparticle textures. For the NTS, the textures are of spherical cap shapes. Detailed analysis of the SEM images shows that the diameters of the textures are approximately between tens of nanometers and about 200 nm, and their heights are approximately 50% of their diameters. Since the diameters of the original nanoparticles in the solution are 18–25 nm, measured by the dynamic light-scattering method, it is believed that small aggregates of the

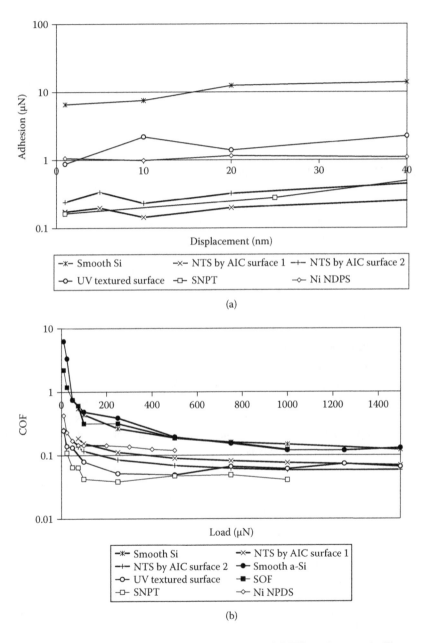

(a)

(b)

FIGURE 6.13 (a) Adhesion force comparison of various MNTSs and a smooth silicon surface. (b) Coefficient of friction comparison of various MNTSs and a smooth Si, a SOF, and
an a-Si surface.

original nanoparticles were melted during annealing and resolidified into spherical caps upon cooling. It has been reported in the literature that nanoparticles have lower melting temperatures than their bulk materials (Peters, Cohen, and Chung 1998; Schmidt et al. 1997; Unruh, Huber, and Huber 1993). For the MTSs, the textures consist of irregularly shaped nanoparticle aggregates. The textures vary from submicron to several microns in lateral dimensions and from tens of nanometers to 100 nm in heights. The percentages of surface covered by the textures obtained using ImageJ (http://rsb.info.nih.gov/ij), an image analysis tool, are 1.9%, 7.3%, and 1.7%, respectively, for the NTS and the MTSs. Statistical analysis of texture size (area) showed that the main difference between the NTS and MTSs is that the MTSs have microscale textures, while the largest diameter of the textures on the NTS is about 200 nm. The texture size ranges and distributions of the NTS and MTSs are also different.

The NTS, MTSs, and a silicon oxide film surface (SOFS) were further characterized by AFM using a silicon tip of less than 10 nm nominal tip radius of curvature in a tapping mode. The results in Figure 6.14 show that the shapes of the textures on the NTS are different from those of the textures on the MTSs. The textures on the NTS have convex shapes at the top of the textures while those of the MTSs have irregular shapes. The radii of curvature of the textures on the NTS in the height direction are much smaller than those on the MTSs. The average maximum peak heights of the NTS and the MTSs based on at least three measurements are 92 nm, 119 nm, and 110 nm, respectively. The heights of the textures measured by the AFM are consistent with the SEM measurement results.

6.3.3.2   Adhesion Performance of the Silica Nanotextured Surface and Silica Microtextured Surfaces

Adhesion tests were performed on the NTS, the MTSs, and a smooth silicon (100) surface using a diamond tip with radius of curvature of 100 μm. Figure 6.15(a) shows a comparison of the average adhesion forces of three measurements performed by the 100-μm tip under increasing maximum indentation displacements on these four surfaces. The adhesion force axis is plotted on a base-10 logarithmic scale for better visualization of the adhesion force differences on all surfaces. Ideally, a SOFS should be used for the comparison. The reason for using a Si(100) surface instead of an SOFS for the comparison was that the adhesion forces for the SOFS were so large that they prevented the tip from lifting off from the SOFS and therefore caused difficulties in making correct adhesion force measurements. The adhesion force of the SOFS is expected to be larger than that of a Si(100) surface because silicon oxide is not conductive and potentially has a larger electrostatic charge buildup, which also contributes to the adhesion force.

It can be seen that the adhesion forces of the NTS and the MTSs are much smaller than those of the Si(100) surface. The adhesion forces of the NTS and the MTSs were only 2–4%, 5–12%, and 3–5%, respectively, of those of the Si(100) surface. In general, the adhesion forces decrease as the percentage of surface coverage decreases. However, at a similar surface coverage, the NTS showed smaller adhesion forces than the MTS. The adhesion data in Figure 6.15(a) can be explained

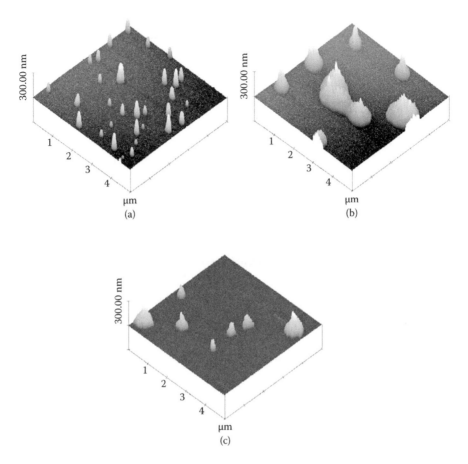

FIGURE 6.14 AFM images of (a) an NTS, (b) an MTS (7.3% surface coverage), and (c) another MTS (1.7% surface coverage) fabricated by spin coating of colloidal silica nanoparticle solution. Data from M. Zou, W. Seale, and H. Wang. Comparison of tribological performances of nano- and micro-textured surfaces. *Proceedings of the Institution of Mechanical Engineers, Part N (Journal of Nanoengineering and Nanosystems)* 219, no. 3 (2005): 103–110.

by the difference between the real areas of contact of the samples. In the case of a diamond tip on the Si(100) surface, the contact area was determined by the radius of curvature of the diamond tip. However, in the case of a diamond tip on the NTS and MTSs, the contact areas were determined primarily by the radii of curvature of the silica nano- or microtextures that are significantly smaller than that of the tip. Therefore, the real contact areas between the tip and the textured surfaces are significantly smaller than the real contact area between the tip and the Si(100) surface, which in turn resulted in significantly smaller adhesion forces on the textured surfaces. It has been shown that the adhesion force between an SPM tip and a flat surface increased almost linearly with increasing SPM tip radius (Yoon et al. 2003). The higher percentage of surface texture coverage, the larger individual

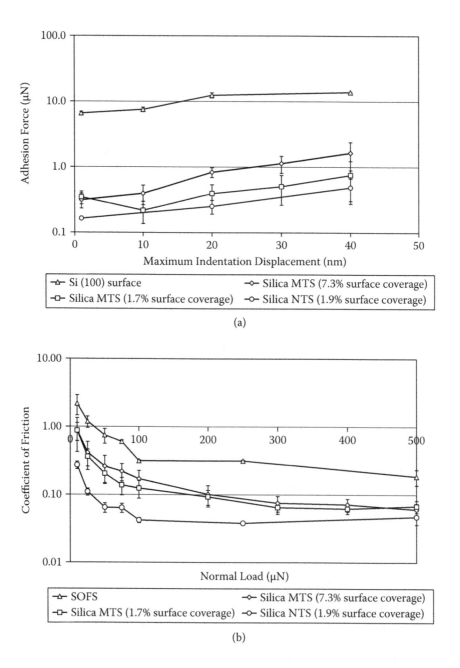

FIGURE 6.15 Adhesion force comparison of the NTS, the MTSs, and a smooth Si(100) surface and (b) COF comparison of the NTS, the MTSs, and the SOFS. Data from M. Zou, W. Seale, and H. Wang. Comparison of tribological performances of nano- and micro-textured surfaces. *Proceedings of the Institution of Mechanical Engineers, Part N (Journal of Nanoengineering and Nanosystems)* 219, no. 3 (2005): 103-10.

texture size, and the concave shapes of the textures are possible reasons for the adhesion forces of the MTS with 7.3% surface coverage to be larger than those of the MTS with 1.7% surface coverage and those of the NTS. The smaller adhesion forces on the NTS compared to those of the MTS with similar surface coverage are because of the smaller real area of contact due to the much smaller radii of curvature of the nanotextures.

6.3.3.3 Friction Performance of the Silica Nanotextured Surface and Silica Microtextured Surfaces

Figure 6.15(b) shows the relationship between the average COFs of three measurements and the normal load for the NTS, the MTSs, and the SOFS measured by the 100-μm tip. The COF axis is plotted on a base-10 logarithmic scale for better visualization of the COF differences on all surfaces at both low and high loads. It can be seen that surface-texturing significantly affected the COFs. The textured surfaces showed significantly reduced COFs compared with the SOFS. For example, at a low load (10 μN), the COFs of the NTS and MTSs are only 12%, 40%, and 40%, respectively, of that of the SOFS. At a high load (500 μN), the COFs of the NTS and MTSs are only 26%, 33%, and 37%, respectively, of that of the SOFS. The NTS has much smaller COFs than those of the MTSs. Comparisons of the surface coverage and COFs of the NTS with those of the MTSs suggest that the reduced surface coverage is not the only reason for the significantly reduced COFs. The reduced local radii of curvatures of individual nanotextures certainly also contributed to this reduction.

Since MEMSs and NEMSs, especially NEMSs, operate at low loads, the impact of surface-texturing at low-load conditions is of particular interest. To understand the physics behind the effect of surface-texturing on the COFs at low loads, the relationship between the COF and load $^{-1/3}$ is plotted in Figure 6.16 for loads less than and equal to 100 μN. Note the linear relationship between the COF and load $^{-1/3}$ for all surfaces. The slopes for the textured surface are much smaller than that of the SOFS, with that of the NTS being the smallest. This suggests that surface-texturing, especially nano-surface-texturing, is very effective in reducing COF at low loads. It is known that the frictional force for an unlubricated surface is proportional to the real contact area and to the shear strength at the asperity junctions (Bowden 1986). Since the real contact area is proportional to load $^{2/3}$ for elastic contact, which is probably the case for all surfaces under low loads, the COF is proportional to load $^{-1/3}$. For the NTS surface at loads greater than or equal to 100 μN, the COFs were nearly independent of the normal load. In these cases, the dominant deformation process is likely to be plastic shearing (plowing) of the nanoparticles on the surfaces by the diamond tip. Examination of the normal displacement data suggest that plastic deformations indeed happened in these cases. When the ploughing mechanism prevails over adhesion, the COF depends primarily on the shape of the hard asperity (diamond tip) and the shear strength (or hardness) of the plowed material (silica-textured surfaces) (Suh and Sin 1981; Komvopoulos, Saka, and Suh 1985; Komvopoulos, Saka, and Suh 1986).

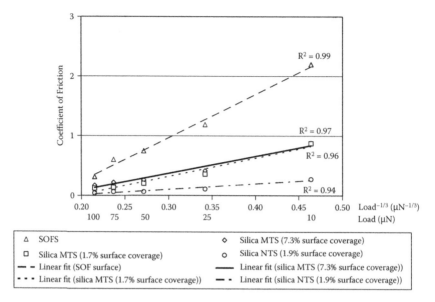

FIGURE 6.16   Linear relationship between coefficient of friction and load $^{-1/3}$. Data from M. Zou, W. Seale, and H. Wang. Comparison of tribological performances of nano- and micro-textured surfaces. *Proceedings of the Institution of Mechanical Engineers, Part N (Journal of Nanoengineering and Nanosystems)* 219, no. 3 (2005): 103–110.

6.4   FUNDAMENTAL UNDERSTANDING OF FRICTION OF MICRO- AND NANOTEXTURED SURFACES

Friction is a phenomenon that has been studied for centuries, since the early days of Leonardo da Vinci (1452–1519). Amontons' law, which states that friction is proportional to the normal load, has described many macroscopic and microscopic nonadhesive sliding contacts surprisingly well (Rabinowicz 1995; Persson and Tosatt 1996). For miniaturized systems such as MEMS/NEMS, however, Amontons' law no longer applies because the adhesion contribution to friction can no longer be neglected due to the large surface-to-volume ratio of MEMS/NEMS structures and the increased surface smoothness (Corwin and de Boer 2004). In fact, adhesion and friction are two of the main issues affecting the reliability of MEMS/NEMS devices involving contacting interfaces (Komvopoulos 2003). Therefore, a fundamental understanding of friction at nanoscale dimensions and controlling friction through nano-surface-engineering are of great scientific and technological significance. In this section, our preliminary understanding of friction of MNTSs, including the frictional behavior of a Ni NDPS, the effects of tip size and surface roughness, applicability of continuum contact mechanics models, as well as the real area of contact, is discussed.

6.4.1   Frictional Behavior of a Ni Nanodot-Patterned Surface

The frictional behavior of a 1-μm diamond tip sliding on the Ni NDPS was studied (Wang et al. 2007). The friction/scratch tests were conducted using a ramp load

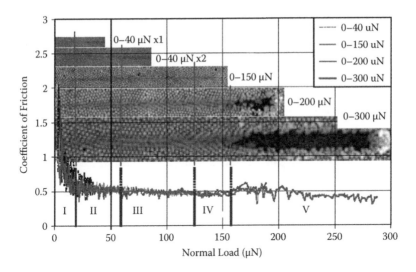

FIGURE 6.17 Coefficient of friction as a function of normal load for four different scratch tests: 0–40 μN, 0–150 μN, 0–200 μN, and 0–300 μN. The inserted SEM micrographs show the corresponding deformations of the sample surfaces after the scratch tests. Hengyu Wang, R. Premachandran Nair, Min Zou, P. R. Larson, A. L. Pollack, K. L. Hobbs, M. B. Johnson, and O. K. Awitor. Friction study of a Ni nanodot-patterned surface. *Tribology Letters* 28 (2007): 183–189, Figure 2. With kind permission from Springer Science+Business Media.

profile from 0 μN to ten different maximum normal loads (10 μN to 500 μN) at a sliding speed of 1 μm/s with 8-μm scratch lengths, as described in Section 6.2.2. Figure 6.17 shows the COF as a function of the normal load for four out of the ten different scratch tests: 0–40 μN, 0–150 μN, 0–200 μN, and 0–300 μN. The four scratch tests were selected to cover both low-load and high-load ranges with sufficient data collection resolution and to show the general trends of the COFs versus normal loads. The inserted SEM micrographs show the corresponding deformations of the sample surfaces after these tests. It can be seen that the COF was not a constant as predicted by Amontons' law. The COF as a function of normal load can be divided into the following five regimes: (I) the COF was relatively high when the normal load was less than 20 μN, (II) the COF approached a constant value of about 0.5 from 20 μN to 60 μN normal load, (III) the COF remained relatively constant at 0.5 until the normal load reached 125 μN, (IV) the COF slightly decreased between 125 μN and 160 μN normal load, and (V) the COF became erratic when the normal load was larger than 160 μN.

The nonconstant behavior of the COF versus normal load can be attributed to different friction mechanisms. At low loads (regime I), the contribution from the adhesion force between the tip and NDPS to the friction force cannot be neglected, as shown in Figure 6.18, which shows the frictional force as a function of the normal load for the four scratch tests presented in Figure 6.17. It can be seen that even though there was a good linear fit when the normal load was less than 125 μN, the fitted line did not go through the origin as predicted by Amontons' law. In other words, the

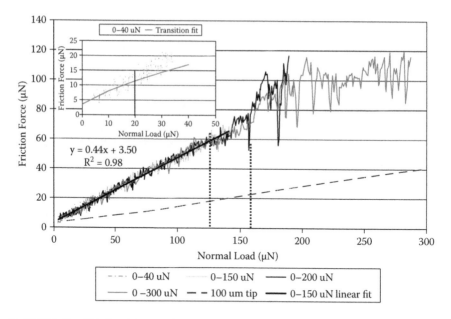

FIGURE 6.18 Frictional force and normal load relationships for four different nominal normal loads. Inset plot shows the frictional force and normal load relationship at low loads fitted with the transition model. Hengyu Wang, R. Premachandran Nair, Min Zou, P. R. Larson, A. L. Pollack, K. L. Hobbs, M. B. Johnson, and O. K. Awitor. Friction study of a Ni nanodot-patterned surface. *Tribology Letters* 28 (2007): 183–189, Figure 3. With kind permission from Springer Science+Business Media.

frictional force was not zero when the normal load was zero. Instead, the frictional force at zero normal load was about 3.5 µN due to the adhesion force.

At higher loads (regime III), Figure 6.17 shows that the COFs were nearly independent of the normal loads. In these cases, the dominant deformation process was plastic shearing (plowing) of the nanodots on the surface by the diamond tip as shown by the inserted SEM micrographs. When the shearing (plowing) mechanism prevails over adhesion, the COF depends primarily on the shape of the hard asperity (diamond tip) and the shear strength (or hardness) of the plowed material (Ni NDPS) (Suh and Sin 1981; Komvopoulos, Saka, and Suh 1985; Komvopoulos, Saka, and Suh 1986); and thus the COF remained constant. Regime II represents a transition regime where both adhesion and plastic deformation mechanisms were present. The inserted SEM images show that in regime IV, the Ni nanodots were sheared severely and started loosening from the substrates. In regime V, the Ni nanodots were completely removed from the substrates and caused the COF to increase suddenly and behave erratically afterward.

The friction-versus-load relationship between a 100-µm tip and the Ni NDPS obtained from the study described in Section 6.3.1.4 (Zou et al. 2006) was also plotted in Figure 6.18 for comparison. It can be seen that the COF between the 1-µm tip and the Ni NDPS (slope of the curve) was about four times the COF between the 100-µm tip and the Ni NDPS. Possible reasons for the much larger COF between the

1-μm tip and the Ni NDPS are: (1) larger plastic deformations due to higher contact pressures at the interface, which resulted from a smaller number of nanodots in contact; (2) interlocking between the 1-μm tip and the Ni nanodots; and (3) larger real area of contact between the 1-μm tip and the NDPS. The fact that the SEM micrograph showed dot flattening under a large range of load suggests that the interlocking is not the major mechanism. The SEM micrographs of the scratches produced by the 1-μm tip showed much smaller deformation track widths than the scratches produced by the 100-μm tip, therefore invalidating the real area of contact argument. Therefore, it was concluded that plastic deformation is the main reason for the large COF of the 1-μm tip. This result suggests that the relative size of the asperities of the mating surfaces affect the COF significantly through the contact area and pressure change. NDPSs can be used to reduce friction only when the asperities of the mating surfaces are relatively large compared to the spacing between the nanodots. This point is further supported by the studies in Section 6.4.2.

6.4.2 EFFECTS OF TIP SIZE AND SURFACE ROUGHNESS ON FRICTIONAL PERFORMANCE

For the MNTSs described in Sections 6.3.1.2, 6.3.1.3, and 6.3.1.4, it was observed that the effect of surface-texturing on the frictional performance was different when the friction tests were performed using a 5-μm tip instead of a 100-μm tip (Zou et al. 2006; Zou, Cai, and Wang 2006; Zou et al. 2005). For example, Figure 6.19 shows the comparison of the friction forces on a nontextured smooth a-Si surface and a textured surface fabricated by UV-assisted crystallization of a-Si measured by a 5-μm tip and a 100-μm tip. It can be seen that, at low loads, surface-texturing has very significant effects on COF measured by both tips. At high loads (>500 μN), the benefit of surface-texturing is still evident from the COFs measured by the 100-μm tip. For the 5-μm tip, however, the COFs of the textured surface are close to the COFs on the nontextured surface areas of 0.09–0.11, which were consistent with the published data for diamond on silicon (Bhushan and Li 1997). It is interesting to notice that the COF on both nontextured and textured surfaces are not constant but change with tip size. As the tip size increased, the COF on the nontextured surface increased, while the COF on the textured surface decreased. This suggests that the COF of the textured surface will continue to decrease with increasing apparent contact area.

To understand the physics behind the effect of the textured surface on COF at low loads, the relationship between the COF and load$^{-1/3}$ measured by the 100-μm tip was plotted for load below 100 μN. The plot shows that there was a nearly linear relationship between the COFs and load$^{-1/3}$ for the textured surface. However, for the nontextured surface, the COFs are almost independent of load$^{-1/3}$. It is known that the friction force for an unlubricated surface is proportional to real contact area and to the shear strength at the asperity junctions (Bowden 1986). Since the real contact area is proportional to load$^{2/3}$ for elastic contact, which is likely the case for the nontextured surface, the COF is proportional to load$^{-1/3}$. For loads greater than or equal to 750 μN, the COF measured by the 5-μm tip was independent of

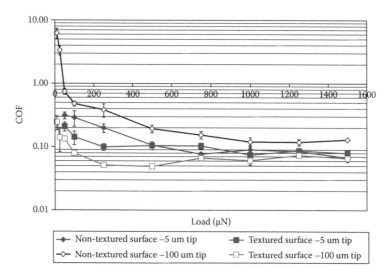

FIGURE 6.19 Coefficient of friction comparison of textured and nontextured surface areas. From M. Zou, L. Cai, H. Wang, D. Yang, and T. Wyrobek. Adhesion and friction studies of a selectively micro/nano-textured surface produced by UV assisted crystallization of amorphous silicon. *Tribology Letters* 20 (2005): 43–52, Figure 10. With kind permission from Springer Science+Business Media.

normal load. In these cases, the dominant deformation process was plastic shearing (plowing) of the surfaces. When the plowing mechanism prevails, the COF depends primarily on the shape of the hard asperity (diamond tip) and the shear strength (or hardness) of the plowed material (silicon surface) (Suh and Sin 1981; Komvopoulos, Saka, and Suh 1985; Komvopoulos, Saka, and Suh 1986). This explains why the COFs of the textured and nontextured surfaces measured by the 5-μm tip are similar at high loads.

To further understand how the surface roughness affects friction, the time histories of the COF, normal displacement, and the slope of the normal displacement with respect to the horizontal displacement on both the textured and nontextured surfaces during sliding contact were investigated. Figure 6.20 shows these relationships measured by a 5-μm tip under 1000 μN normal load. Figure 6.20(a) shows that the variation of the COF of the textured surface is quite large, and it follows the slope rather than the normal displacement. This can be explained by the ratchet mechanism of friction, which states that the friction force is proportional to the slope for small slope angles (Bowden 1986). The maximum slope angle for this test condition was less than 3°, which satisfies the small slope angle condition. Figure 6.20(b), on the other hand, shows there was little variation of the COF on the nontextured surface because the topography is smooth and the slope of the surface hardly changed. Therefore, the variation of the COF with slope only happened when the tip was small enough to follow the topography of the surface to some extent. When the tip was relatively large compared to the micro- or nanotextures, as in the case of the 100-μm tip, the COF no longer showed variations during sliding.

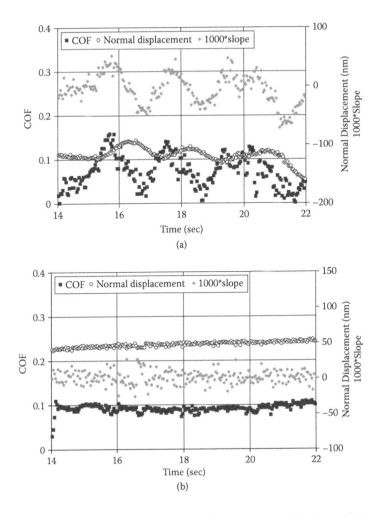

FIGURE 6.20 Coefficient of friction, normal displacement, and the slope of the surface topography vs. time on (a) the textured surface and (b) the nontextured surface during sliding contact under 1000 μN normal load measured by a 5-μm tip. From M. Zou, L. Cai, H. Wang, D. Yang, and T. Wyrobek. Adhesion and friction studies of a selectively micro/nano-textured surface produced by UV assisted crystallization of amorphous silicon. *Tribology Letters* 20 (2005): 43–52, Figure 12. With kind permission from Springer Science+Business Media.

6.4.3 APPLICABILITY OF CONTINUUM CONTACT MECHANICS MODELS

Continuum contact mechanics models have been applied to nanotribology measurements to determine fundamental parameters and processes in nanometer-scale single-asperity contacts for the past 15 years (Grierson, Flater, and Carpick 2005). An increasing amount of data supports the surprising conclusion that a continuum description of contact is sometimes accurate down to nanometer-sized single-asperity contacts. This is particularly surprising because many of the basic assumptions associated with these continuum contact mechanics models, such as homogeneous, isotropic, and linear elastic materials, and the contact radius being much smaller than the radius of curvature of the contacting interfaces, may be violated for some of the interfaces studied. Several adhesive continuum contact mechanics theories have been developed based on Hertz theory (Johnson 1987). These theories consider attractive forces between the contacting asperities. The JKR theory (Johnson, Kendall, and Roberts 1971) and the DMT theory (Derjaguin, Muller, and Toporov 1975) represent opposite ends of the spectrum of a nondimensional transition parameter (representing the ratio of the normal elastic deformation caused by adhesion to the effective range of the adhesion forces). Generalization of these two limiting cases to an intermediate case applicable to an actual interface interaction is achieved by Maugis using a Dugdale model in continuum fracture mechanics (Maugis 1992), and further simplified by Carpick et al. (1999) using a generalized transition model.

Theories of continuum contact mechanics were employed to explain the relationship between the friction force and the applied normal load for the tip–Si(100) and tip–Ni NDPS interfaces described in Section 6.3.1.4 (Zou et al. 2006). In Figure 6.21, the friction force versus normal load data for both the tip–Si(100) and tip–Ni NDPS interfaces is plotted along with the fitted data for the low-load (elastic contact) regime using continuum contact mechanics models according to a procedure described by Grierson et al. (Grierson, Flater, and Carpick 2005). First, the appropriate continuum contact models to fit the low-load regime of the friction force versus normal load curves were determined from the measured average adhesion force at 1-nm indentations for both the tip–Si(100) interface and the tip–Ni dot interface (the adhesion force was divided by 80, the estimated number of Ni dots in

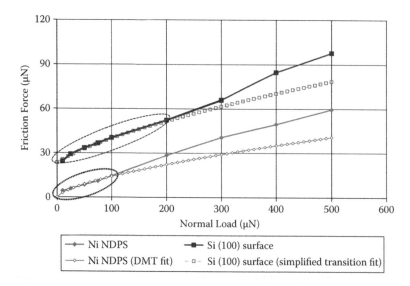

FIGURE 6.21 Friction force as a function of applied normal load and continuum contact mechanics models that fit the data. Hengyu Wang, R. Premachandran Nair, Min Zou, P. R. Larson, A. L. Pollack, K. L. Hobbs, M. B. Johnson, and O. K. Awitor. Friction study of a Ni nanodot-patterned surface. *Tribology Letters* 28 (2007): 183–189, Figure 6. With kind permission from Springer Science+Business Media.

contact). We assumed the Young's modulus for Si(100), Ni dot, and diamond to be 161 GPa, 200 GPa, and 1141 GPa, respectively, and the Poisson's ratio for Si(100), Ni dot, and diamond as 0.23, 0.31, and 0.07, respectively. We also assumed the value of the effective range of adhesion for the Si(100) surface and the Ni NDPS to be 0.249 nm and 0.234 nm, which are the Si–Si and Ni–Ni bond distances, respectively. The transition parameters obtained from these values are 0.41 for the Si(100) surface and 0.061 for the Ni NDPS. Therefore, the tip–Si(100) interface is in the transition regime, and the tip–Ni NDPS is in the DMT regime. Second, we fit the friction force versus normal load curves using the generalized transition model (Carpick, Ogletree, and Salmeron 1999) and the DMT model (Derjaguin, Muller, and Toporov 1975), respectively.

From Figure 6.21, it can be seen that for the Si(100) surface, the generalized transition model fits the experimental data very well under loads less than 200 µN, suggesting the adhesion force contribution to the friction force is significant at loads less than 200 µN. The fact that the elastic contact model applies to the experimental data also suggests that the contact deformation was in the elastic regime. For the Ni NDPS, the DMT contact model fits the experimental data reasonably well for normal loads below 100 µN, suggesting the adhesion force contributes to the friction force at loads less than 100 µN. The transition model/DMT model did not fit the data for loads greater than 200 µN/100 µN, suggesting the possibility that plastic deformation occurs at normal loads larger than 200 µN for the tip–Si(100), and 100 µN for the tip–Ni NDPS interfaces, respectively.

The applicability of a continuum contact mechanics model to the nanoscale multi-asperity contact between the diamond tip and the Ni NDPS is very intriguing. When the tip is in contact with the Ni NDPS, contact force is supported by multiple Ni nanodots of the same size. The elastic contact model applies to each Ni dot–tip interface if the contact interface materials are homogeneous and only elastic deformation occurs. Since the Ni NDPS is a combination of Ni nanodot and Si substrate, and therefore is inhomogeneous, it is very surprising to see that the classical contact mechanics model can still be used to explain this nanoscale multi-asperity contact for such an inhomogeneous material.

For the Ni NDPS at loads greater or equal to 400 µN, Figure 6.12(b) showed that the COFs were nearly independent of the normal load. In these cases, the dominant deformation process is likely to be plastic shearing (plowing) of the nanodots on the surfaces by the diamond tip. Examination of the normal displacement data suggests that plastic deformation indeed occurred in these cases. When the plowing mechanism prevails over adhesion, the COF depends primarily on the shape of the hard asperity (diamond tip) and the shear strength (or hardness) of the plowed material (Ni NDPS) (Suh and Sin 1981; Komvopoulos, Saka, and Suh 1985; Komvopoulos, Saka, and Suh 1986).

A continuum contact mechanics model was also employed to quantify the relationship between the frictional force measured by a 1-µm diamond tip on the Ni NDPS and the applied normal load described in Section 6.4.1 for loads smaller than 40 µN (Wang et al. 2007). In the inset in Figure 6.18, the frictional force versus normal load data was plotted for normal loads less than 40 µN. Also plotted was the fitted data using a continuum contact mechanics model according to a procedure described by Grierson et al. (Grierson, Flater, and Carpick 2005). First, the appropriate continuum contact model was determined from the measured average adhesion force at 10-μN indentations. We took the value of the Young's modulus for Ni and diamond as 200 GPa and 1141 GPa, respectively, and the Poisson's ratio for Ni and diamond as 0.31 and 0.07, respectively. We also assumed the value of the effective range of adhesion for the Ni NDPS to be 0.234 nm, which is the Ni–Ni bond distance. The transition parameter obtained from these values is 0.64, which is larger than 0.1 and less than 5. Therefore, the tip–Ni NDPS interface was determined to be in the transition regime. Second, we fit the frictional force versus normal load curves using the generalized transition model (Carpick, Ogletree, and Salmeron 1999). From the inset in Figure 6.18, it can be seen that the generalized transition model fits the experimental data very well for normal loads less than 20 μN, again suggesting the adhesion force contribution to the frictional force is significant at loads less than 20 μN.

The elastic contact model applies to each Ni dot–tip interface if the contact interface materials are homogeneous and only elastic deformation occurs. Since the Ni NDPS (a combination of nanometer-sized Ni dots and the Si substrate) is inhomogeneous and plastic deformation occurred at low loads in this study (shown later in Section 6.4.4), it is again very interesting to see that a continuum contact mechanics model can still be used to explain the nanoscale contact. The reason for this is not fully understood yet and will be the subject of future study.

6.4.4 Deformation of a Ni Nanodot-Patterned Surface and Real Area of Contact

Detailed SEM investigations were conducted to reveal the deformation of the Ni NDPS after scratching using a 1-μm diamond tip at different normal loads (Wang et al. 2007). Figure 6.22 shows an example of SEM analysis on a low-load 0–40 μN scratch. Four SEM micrographs with different magnifications were presented. Figure 6.22(a) shows the unprocessed SEM micrograph, on which the scratch can hardly be seen. However, after adjusting the brightness and contrast of the original SEM micrograph, the scratch is clearly shown in Figure 6.22(b). To show the Ni nanodot deformation better, SEM micrographs were taken at higher magnifications, and are shown in Figure 6.22(c) and (d). It can be seen from Figure 6.22(c) that the Ni nanodots deformed plastically even at very small normal loads at the beginning of the scratch. The plastic deformation of the Ni nanodots at loads greater than 20 μN was clearly shown by the altered shapes of the dots. The real area of contact, shown in Figure 6.22(b) between the dotted lines, increased with the applied normal load. Since Figure 6.18 shows that the frictional force was proportional to the normal load when the normal load was less than 40 μN, we concluded that the frictional force was proportional to the real area of contact, the same as generally assumed for macroscale friction.

AFM measurements were taken to further quantify the deformation of the Ni NDPS in the height direction under different normal loads. Results showed that at the beginning of the scratch, only one or two Ni nanodots were in contact with the tip

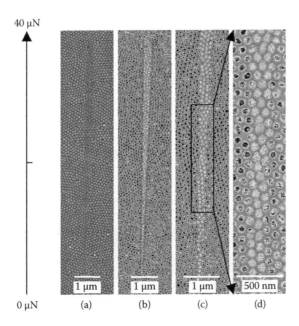

FIGURE 6.22 SEM micrographs illustrating the deformations of the Ni dots after a 0–40 µN scratch test. (a) Unprocessed micrograph, (b) adjusted micrograph brightness and contrast to show the scratch better, (c) and (d) higher magnification micrographs. Hengyu Wang, R. Premachandran Nair, Min Zou, P. R. Larson, A. L. Pollack, K. L. Hobbs, M. B. Johnson, and O. K. Awitor. Friction study of a Ni nanodot-patterned surface. *Tribology Letters* 28 (2007): 183–189, Figure 4. With kind permission from Springer Science+Business Media.

and the scratch test caused the Ni nanodots to deform plastically in the whole length of the scratch. Cross-sectional line scans revealed that even at the very beginning of the scratch, the Ni nanodots were deformed about 5–10 nm in the height direction. The estimated maximum shear stress for a single asperity contact under 4-µN normal load, which was the actual normal load applied at the beginning of the scratch instead of 0 µN specified due to the imprecision of the TriboIndenter, was 8.4 GPa assuming the Hertz contact model without lateral force (Johnson 1987). This shear stress is much larger than the theoretical shear strength of Ni of 2.6 GPa (Courtney 2000), which therefore caused the Ni dot to deform readily. However, the relatively small deformation under such a high contact pressure indicates the unusual strength of the Ni nanodots.

The frictional force versus lateral displacement curve contains a wealth of information regarding the deformation behavior of the NDPS and the critical shear strength of the Ni nanodots–Si substrate interface. Figure 6.23 shows the relationship between the frictional force and the lateral displacement for a 0–200 µN scratch, along with the SEM micrograph of the surface after the scratch. It can be seen that there is a very good correlation between the frictional force and the surface deformation. The five regimes identified in Figure 6.17 can also be seen in Figure 6.23 with the aid of the SEM micrograph, which provided evidence of deformation in each regime. The SEM micrograph shows only slight plastic deformation

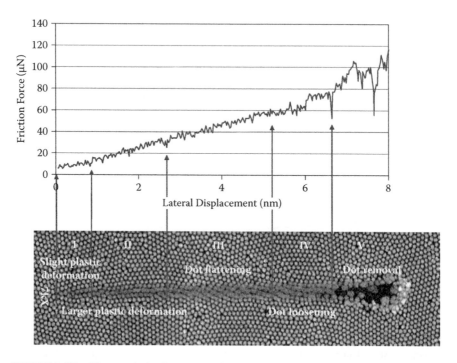

FIGURE 6.23 The correlation between surface topography and the signatures in the lateral force vs. lateral displacement curve for a 0–200 μN scratch. Hengyu Wang, R. Premachandran Nair, Min Zou, P. R. Larson, A. L. Pollack, K. L. Hobbs, M. B. Johnson, and O. K. Awitor. Friction study of a Ni nanodot-patterned surface. *Tribology Letters* 28 (2007): 183–189, Figure 6. With kind permission from Springer Science+Business Media.

of Ni nanodots in regime I and transition to a larger plastic deformation in regime II. In regime III, significant flattening of dots was observed. Finally, in regime IV, the dots became loose, and they were completely removed in regime V. The frictional force versus lateral displacement curve showed clear indications of transitions from one regime to the other.

The critical shear strength of the Ni nanodots–Si substrate interface was estimated from the shear stress at which the Ni nanodots were removed from the silicon substrate. By examining the lateral force and SEM micrographs of the scratch together, we found that the critical shear stress point corresponds to the point at which the lateral force becomes irregular (critical lateral force). Therefore, the critical shear stress of the Ni nanodots–Si substrate interface was calculated by dividing the critical lateral force by the area of contact between the tip and the NDPS at the critical lateral force determined from the SEM micrographs of the scratch. The critical shear stress at the Ni nanodots removal for 9 different scratches gave an average of 1.24 GPa and a standard deviation of 0.01 GPa. This shear stress is very close to the reported value of 1.4 GPa for the Ni–silica interface (Agrawal and Raj 1990).

6.5   EFFECTS OF SURFACE WETTING PROPERTY ON THE ADHESION AND FRICTION PERFORMANCES OF MICRO- AND NANOTEXTURED SURFACES

Large adhesion and friction forces in MEMS applications can be alleviated by either increasing surface roughness through surface-texturing or improving surface hydrophobicity through applying low-surface-energy materials to MEMS surfaces (Komvopoulos 2003; Maboudian and Howe 1997). Surface-texturing can reduce the contact area and thus decrease adhesion and friction forces between surfaces. Improving surface hydrophobicity can reduce water adsorption on the surface and therefore reduce meniscus-mediated adhesion and friction forces between contacting surfaces. Since surface hydrophobicity significantly affects surface adhesion and friction properties, we focused our efforts on producing superhydrophobic/hydrophobic surfaces and characterizing the wetting, adhesion, and friction properties of such surfaces.

6.5.1   EFFECTS OF ROUGHNESS ON SURFACE WETTING

The surface wetting property is directly reflected by the WCA of a surface. Chemically modifying smooth silicon surfaces can only lead to hydrophobic surfaces having WCAs of up to 120° (Tambe and Bhushan 2005). To achieve superhydrophobicity of a surface, textured surfaces are necessary. The topography effect on WCA can be explained by either the Wenzel model (Wenzel 1936) or the Cassie-Baxter model (Cassie and Baxter 1944). The latter assumes that a water droplet contacts the top of a rough surface with air trapped between the rough asperities on the surface. The apparent WCA, θ^*, is given by

$$\cos\theta^* = f_1 \cos\theta - f_2 \tag{6.1}$$

where θ is the WCA of a smooth surface with the same composition as the rough surface, f_1 is the surface area fraction of the solid, f_2 is the surface area fraction that are voids, and $f_1 + f_2 = 1$. Thus, with proper surface-texturing and surface chemical modification, it is possible to produce a surface having a θ^* much larger than θ (Martines et al. 2005; Feng et al. 2002; Ming et al. 2005).

We have developed a MEMS process-compatible method for generating superhydrophobic surfaces for the purpose of reducing adhesion and friction forces in MEMS. The method includes using rapid AIC of a-Si technique to produce silicon MNTSs on silicon substrates and applying octadecyltrichlorosilane (OTS) self-assembled monolayers (SAMs) on the textured surfaces to reduce the surface energies of the textured surfaces.

The process of using rapid AIC of a-Si to produce MNTSs was described in Section 6.2.1.1. To produce MNTSs with different surface topographies, a-Si films of 100 nm to 400 nm were deposited followed by deposition of 800-nm aluminum film. The samples were then cut into small pieces of 1 inch × 1 inch and annealed in air in a conventional furnace at various temperatures ranging from 750°C to 850°C with durations varying from 5 s to 20 s (Song et al. 2010).

The MNTSs, along with thermally oxidized smooth silicon samples (control samples), were then soaked in piranha solution ($H_2O_2 : H_2SO_4 = 3 : 7$ by volume) at room temperature for an hour to remove contaminants and grow a fresh oxide layer on the substrate to facilitate the assembly of OTS on the surfaces. After piranha cleaning, the samples were rinsed with deionized (DI) water and toluene and blow-dried with N_2 gas. The samples were then dipped into an OTS-toluene solution with an OTS mass concentration of 1% for 10 min to allow the OTS to uniformly self-assemble on the sample surfaces. Finally, the samples were rinsed with chloroform ($CHCl_3$), cleaned ultrasonically for 5 min in chloroform, rinsed again with chloroform, and then dried in air. The above OTS deposition conditions on MNTSs are the optimized deposition conditions obtained from our previous studies (Song et al. 2009).

Figure 6.24 shows representative SEM micrographs of an oxidized silicon wafer and a nanotextured sample. Figure 6.24(a) shows that the oxidized silicon wafer surface is smooth. Figure 6.24(b) shows that, after the AIC of a-Si process, the surface of the silicon wafer was textured by irregularly shaped nanoscale islands of various sizes. Higher-magnification SEM images, taken from the textured surface at 70° tilt

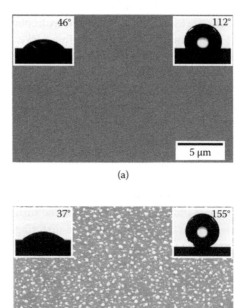

(a)

(b)

FIGURE 6.24   SEM micrographs of (a) a smooth oxidized silicon surface and (b) a nanotextured surface. Insets of (a) are WCA pictures of the smooth oxidized silicon surface before (left) and after (right) OTS deposition. Insets of (b) are WCA pictures of the nanotextured surface before (left) and after (right) OTS deposition.

angle, show that irregularly shaped textures are randomly distributed on the surface, with sizes ranging from 10s of nanometers to several hundred nanometers in both lateral and height directions. SEM micrographs of oxidized smooth silicon surfaces and MNTSs after OTS SAM deposition show no aggregated OTS islands and no discernable surface topography change after the OTS self-assembly on both the smooth and the textured surfaces. In other words, OTS SAM did not change surface topography of the smooth or the textured surfaces.

The surface wetting property of the oxidized smooth silicon surfaces and the MNTSs shown in Figures 6.24(a) and (b), before and after OTS deposition, were characterized by a video-based contact angle measurement system. The insets in Figure 6.24(a) show that the WCA of the oxidized smooth silicon sample was 46° before OTS deposition and 112° after OTS deposition. The 112° WCA matches the reported WCA of a smooth Si surface fully covered by OTS (Mirji 2006). The insets in Figure 6.24(b) show that the WCA of the textured sample was 37° and 155° before and after OTS deposition, respectively. Further characterization shows that the water sliding angle of the OTS-modified MNTSs is smaller than 1°. The WCA of the OTS-modified textured surface (155°) was much higher than that of the OTS SAM modified smooth surface (112°), confirming the importance of surface texturing on improving the surface hydrophobicity.

The superhydrophobicity of the OTS-modified textured Si surfaces can be explained based on both chemical and topographic factors. Before OTS self-assembly, the terminal groups of the oxidized Si surface were O or OH, resulting in a hydrophilic surface. After the OTS self-assembly was complete, the hydrophilic groups were no longer available, and the surface became hydrophobic because of the hydrophobic alkyl chain of the OTS. Surface-texturing further enhances the hydrophobicity of the surface because when a water droplet is dropped on a textured surface, air is likely to be trapped between the textures and the water droplet, which prevents the water droplet from wetting the surface between the textures.

Because surface topography has significant effects on surface wetting properties, such effects were further studied by investigating the wetting properties of various topographies. Four groups of samples A, B, C, and D, with different topographies, were produced by changing a-Si thickness and annealing conditions. Samples from group A were produced with 100-nm a-Si, annealed at 850°C for 5 s; samples from group B were produced with 300-nm a-Si, annealed at 750°C for 10 s; samples from group C were produced with 200-nm a-Si, annealed at 800°C for 20 s; and samples from group D were produced with 400-nm a-Si, annealed at 800°C for 15 s. The reason for selecting these process parameters is that they produced four distinct topographies.

Figure 6.25 shows representative SEM micrographs of the MNTSs from the four groups after OTS modification. The insets at the upper right corners of the SEM images are the WCA images taken from the corresponding samples, and the insets next to the WCAs images are the higher magnification SEM images. It can be seen that the four samples have significantly different surface topographies and surface texture area coverage. Figures 6.25(a), (b), and (c) show that the surfaces of samples A, B, and C were covered by isolated polycrystalline silicon nano- and microislands. The nano- and micro-sized silicon crystallites are randomly distributed on the surfaces, forming nano- and microtextured silicon surfaces. Figure 6.25(d) shows that

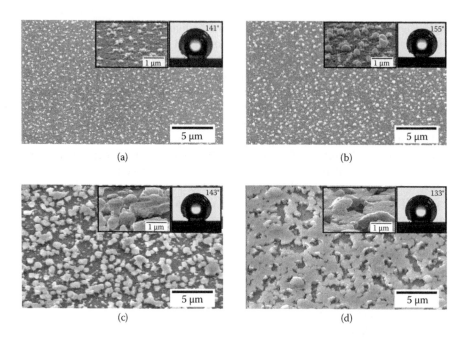

FIGURE 6.25 SEM micrographs of four textured samples of different topographies after OTS deposition. The left insets are higher-magnification SEM images of the sample surfaces, taken at 70° tilt angle. The right insets are optical images of a water droplet on the surfaces. (a) 100 nm a-Si, 10% coverage, WCA = 141°, (b) 300 nm a-Si, 15% coverage, WCA = 155°, (c) 200 nm a-Si, 34% coverage, WCA = 143°, and (d) 400 nm a-Si, 50% coverage, WCA = 133°. Data from Y. Song, R. Premachandran Nair, M. Zou, and Y. A. Wang. Adhesion and friction properties of micro/nano-engineered superhydrophobic/hydrophobic surfaces. *Thin Solid Films* 518, no. 14 (2010): 3801–3807.

the surface of sample D was covered by networks of large silicon grains protruding from the substrate surface.

The WCAs of the OTS-modified textured samples shown in the upper right corners in Figure 6.25 are all much larger than that of the OTS-modified smooth surface, with the largest WCA of 155°. This can be explained by the Cassie-Baxter model (Cassie and Baxter 1944), which describes the wetting of rough surfaces. Using the WCA value of an OTS-modified smooth silicon surface of 112° and the surface area fraction of the solid of 10%, 15%, 34%, and 55%, for the four textured surfaces, respectively, the Cassie-Baxter model predicts the WCA of OTS-modified nano- and microtextured silicon surfaces to be 160°, 155°, 142°, and 131° for the samples shown in Figures 6.25(a), (b), (c), and (d), respectively. The surface area fraction of the solid was calculated using ImageJ by selecting the highest level of textures. These WCA predictions are in good agreement with the experimentally measured WCAs, with the exception of the sample shown in Figure 6.25(a). Because this sample has very short and sparse textures, as shown in Figure 6.25(a) left inset, it is not likely for the water droplet to just contact the textures without touching the rest of the substrate surface. Therefore, the Cassie-Baxter model does not apply to this sample.

6.5.2 Correlation among Surface Wetting, Adhesion, and Friction

Figure 6.26(a) shows a comparison of the adhesion force (in logarithmic scale) versus indentation displacement of a smooth and a textured sample before and after OTS modification. It can be seen that even though a tip with much smaller radius (5 µm) was used for the adhesion test of the smooth Si sample, the adhesion force of the smooth Si surface is still much higher than that of the other three samples (one nanotextured sample, one OTS-modified smooth Si sample, and one OTS-modified nanotextured sample) tested by a 100-µm tip, which indicates that both surface-texturing and OTS modification reduce adhesion forces.

The reduction of adhesion forces on the textured surfaces and the OTS-modified smooth surfaces can be understood through the following reasoning. It has been shown that the adhesion force between an SPM tip and a flat surface increases almost linearly with the SPM tip radius (Yoon et al. 2003). In the case of a diamond tip on the smooth surface, the contact area was solely determined by radius of curvature of the diamond tip. However, in the case of a diamond tip on the textured surface, the contact area was determined not only by the radius of curvature of the diamond tip, but also by the size of micro- and nanoislands that are significantly smaller than the radius of curvature of the tip. Therefore, the real contact areas between the diamond tip and the textured surfaces are significantly smaller than the real contact area between the diamond tip and the flat surfaces, which results in smaller adhesion forces. The hydrophobic nature of the OTS SAMs may cause fewer water molecules to adsorb on the surface, which in turn could reduce the adhesion force caused by water menisci.

Comparison of adhesion forces on the textured surfaces and those on the OTS-modified smooth Si surface shows that surface texturing reduced the adhesion forces more than the OTS modification, possibly due to larger surface separation and less chance of meniscus formation. Among all the investigated samples, the OTS-modified textured surface showed the lowest adhesion due to the benefit of both surface texturing and OTS modification.

Figure 6.26(b) shows the COF versus normal load of the above four samples. Similar to the adhesion results, the smooth Si sample showed the largest COF. However, the OTS-modified smooth Si showed the lowest COF. One possible reason is that, when fully assembled, the molecules of OTS are closely packed and oriented nearly perpendicular to the surface (Mirji 2006); during sliding, the densely assembled OTS molecules tilted in the same direction. The tilted orientation reduced the interfacial shear strength between the tip and the sample surface, and hence lowered friction (Bhushan and Liu 2001). Additionally, the low surface energy of the OTS and the lower chance of meniscus formation on surfaces are also likely to reduce the adhesion contribution to friction. Figure 6.26(b) also shows that the two textured surfaces with and without OTS modification had smaller COF than the smooth surface, mainly because of the much reduced contact area between the tip and the surface due to the surface texturing. However, there was no significant difference in the COFs between these two textured surfaces. This is due to the lack of OTS ordering on textured surfaces. As a result, OTS modification was not effective in further reduction of friction beyond what the surface texturing can do.

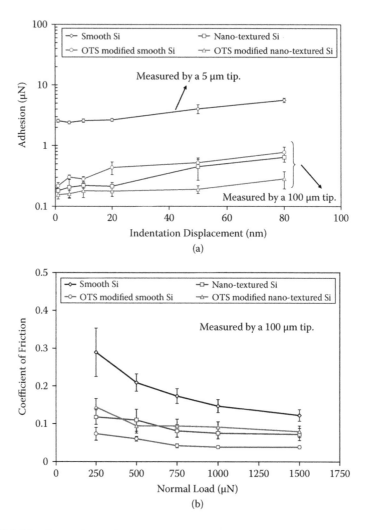

FIGURE 6.26   Comparisons of (a) adhesion and (b) friction properties of a smooth and a nanotextured sample before and after OTS modifications. Data from Y. Song, R. Premachandran Nair, M. Zou, and Y. A. Wang. Adhesion and friction properties of micro/nano-engineered superhydrophobic/hydrophobic surfaces. *Thin Solid Films* 518, no. 14 (2010): 3801–3807.

The effects of surface topography on the adhesion and friction properties of the micro- and nanoengineered surfaces were investigated. Figure 6.27(a) shows, once again, that the adhesion forces of the OTS-modified surfaces measured with a 100-μm tip are significantly smaller than those of the smooth silicon surface measured with a 5-μm tip. It also suggests a strong correlation between adhesion performance and surface wetting properties. The sample with higher WCA showed lower adhesion forces. The best adhesion performance was obtained on the surface with the largest WCA. At an indentation depth up to 50 nm, the adhesion force difference between the samples made from 100-nm a-Si (Sample A) and

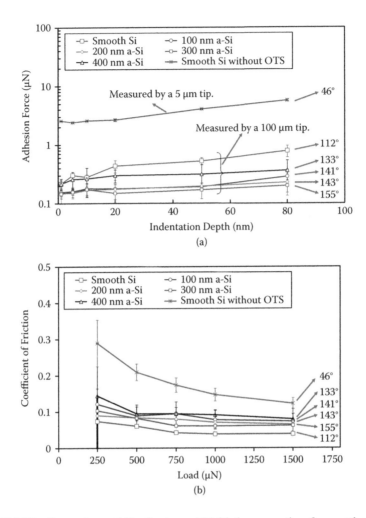

FIGURE 6.27 Comparisons of (a) adhesion and (b) friction properties of a smooth and various micro- and nanotextured samples after OTS SAM modifications. Data from Y. Song, R. Premachandran Nair, M. Zou, and Y. A. Wang. Adhesion and friction properties of micro/nano-engineered superhydrophobic/hydrophobic surfaces. *Thin Solid Films* 518, no. 14 (2010): 3801–3807.

200 nm (Sample C) was minimal because the two samples had about the same WCAs. At an 80-nm indentation depth, the adhesion force of the sample made from 100-nm a-Si (Sample A) was slightly larger than that of the sample made from 200-nm a-Si (Sample C) due to the shorter distance between the tip and the substrate resulting from larger deformations of the relatively smaller textures and the originally shorter texture height. The samples made from 400-nm a-Si (Sample D) showed larger adhesion force than the other three textured surfaces (Sample A, B, and C) due to its smaller WCA and larger surface texture area coverage. The

OTS-modified smooth silicon sample showed the largest adhesion force among the OTS-modified samples.

The frictional performance of the OTS-modified samples was significantly improved compared to smooth Si surfaces as shown in Figure 6.27(b). Similar to results presented in Figure 6.26(b), the OTS-modified smooth silicon had the lowest COF. This, again, is possibly due to the densely assembled OTS molecules tilted in the same direction during sliding, which reduced the interfacial shear strength between the tip and the sample surface, and thus the COF. Additionally, the low surface energy of the OTS SAM and the lower chance of meniscus formation on the surfaces are also likely to reduce the adhesion contribution to friction. For the OTS-modified textured surfaces, the surface textures reduced the contact area, and thus reduced the COF. However, as noted earlier, due to the curvature caused by the textures, OTS molecules are likely to assemble in random orientations rather than in an orderly manner. Therefore, OTS modification on the textured surfaces was not as effective in reducing the friction force as OTS modification on a smooth silicon surface. Figure 6.27(b) shows that, for the textured surfaces only, the wetting property also correlates with their COF; that is, the larger the WCA of a textured sample, the lower its COF. The measured deformation in the vertical direction during sliding showed that the textures were deformed, but there was no indication of texture removal, suggesting the textures can support high contact pressures.

6.6 SUMMARY AND FUTURE DIRECTIONS

Various micro- and nanotexturing methods targeted for solving MEMS/NEMS tribological issues were developed. The textured surfaces were shown to significantly reduce both adhesion and friction forces due to the significantly reduced contact area. Adhesion was found to play a major role in the frictional performance of MNTSs when the normal load was small, while plastic deformation dominated frictional performance when the normal load was large. Surprisingly, continuum contact mechanics models were found to be applicable to the nanoscale contact between a spherical tip and an inhomogeneous Ni NDPS at low loads. The COF was found to be dependent on the size of the tip and surface roughness. Similar to what was assumed at macroscale, the frictional force was found to be proportional to the real area of contact at the nanoscale. By combining surface micro- and nanotexturing with chemical modification using low surface-energy self-assembled monolayers, surface hydrophobicity can be significantly increased. Surface hydrophobicity was found to be directly correlated to the adhesion and friction performance of MNTSs.

As devices continue to shrink in size, NTSs become important for NEMS applications. Even though many methods have been investigated for fabricating NTSs, creating large-scale NTSs economically is still a major roadblock for commercializing NTSs for tribological applications. The majority of current NTS fabrication methods, such as nanoimprinting lithography and soft lithography, are mainly applicable to polymer materials. These materials deform and wear easily and thus are less desirable for tribological applications. It is more desirable to have nanotextures that are mechanically strong and wear-resistant. Furthermore, the adhesion between nanotextures and substrates also needs to be strong to ensure the nanotextures will not be removed easily. In addition, integrating surface nanotexturing processes into existing device fabrication may also be challenging. All these aspects will be the subjects of future investigations.

ACKNOWLEDGMENTS

Financial support for the work described in this chapter was provided by the National Science Foundation, Arkansas Biosciences Institute, and the Oak Ridge Associated Universities. Contributions from my students Hengyu Wang, Ying Song, and Rahul Premachandran Nair, and collaborators Drs. Mathew Johnson (University of Oklahoma), Andrew Wang (Ocean NanoTech, LLC.), and Dehua Yang (Ebatco) are gratefully acknowledged.

REFERENCES

Agrawal, D. C., and Raj, R. 1990. Ultimate shear strengths of copper-silica and nickel-silica interfaces. *Materials Science and Engineering A* 126: 125–131.

Alley, R. L., et al. 1992. *The effect of release-etch processing on surface microstructure stiction*. New York: IEEE.

Bhushan, B., and Huiwen Liu. 2001. Nanotribological properties and mechanisms of alkylthiol and biphenyl thiol self-assembled monolayers studied by AFM. *Physical Review B (Condensed Matter and Materials Physics)* 63(24): 245412.

Bhushan, B., and Xiaodong Li. 1997. Micromechanical and tribological characterization of doped single-crystal silicon and polysilicon films for microelectromechanical systems devices. *Journal of Materials Research* 12(1): 54–63.

Bowden, F. P. 1986. *The friction and lubrication of solids*. Oxford, UK: Clarendon.

Carpick, Robert W., D. F. Ogletree, and Miquel Salmeron. 1999. A general equation for fitting contact area and friction vs. load measurements. *Journal of Colloid and Interface Science* 211(2): 395–400.

Cassie, A. B. D., and S. Baxter. 1944. Wettability of porous surfaces. *Transactions of the Faraday Society* 40: 546–551.

Chik, H., et al. 2004. Periodic array of uniform ZnO nanorods by second-order self-assembly. *Applied Physics Letters* 84(17): 3376–3378.

Chilamakuri, S. K., and B. Bhushan. 1997. Optimization of asperities for laser-textured magnetic disk surfaces. *Tribology Transactions* 40(2): 303 –311.

Choi, Chang-Hwan, Joonwon Kim, and Chang-Jin Kim. 2004. *Nanoturf surfaces for reduction of liquid flow drag in microchannels*. Pasadena, CA: American Society of Mechanical Engineers.

Choi, Dae-Geun, Se G. Jang, Hyung K. Yu, and Seung-Man Yang. 2004. Two-dimensional polymer nanopattern by using particle-assisted soft lithography. *Chemistry of Materials* 16(18): 3410–3413.

Corwin, A. D., and M. P. de Boer. 2004. Effect of adhesion on dynamic and static friction in surface micromachining. *Applied Physics Letters* 84(13): 2451–2453.

Courtney, T. H. 2000. *Mechanical Behavior of Materials*. 2nd ed. New York: McGraw-Hill.

Derjaguin, B. V., V. M. Muller, and Yu P. Toporov. 1975. Effect of contact deformations on the adhesion of particles. *Journal of Colloid and Interface Science* 53(2): 314–326.

Di Fabrizio, Enzo, et al. 2003. *Nano-optical elements fabricated by e-beam and x-ray lithography*. Proceedings of SPIE 5225 (113), doi:10.1117/12.507880.

Donthu, S. K., et al. 2005. Near-field scanning optical microscopy of ZnO nanopatterns fabricated by micromolding in capillaries. *Journal of Applied Physics* 98(2): 1–5.

Etsion, Izhak. 2005. State of the art in laser surface texturing. *Journal of Tribology* 127(1): 248–253.

Feng, Lin, et al. 2002. Super-hydrophobic surfaces: From natural to artificial. *Advanced Materials* 14(24): 1857–1860.

Fu-Ken Liu, et al. 2003. Rapid fabrication of high quality self-assembled nanometer gold particles by spin coating method. *Mechanical Engineering* 67–68: 702–709.

Gall, S., et al. 2002. Aluminum-induced crystallization of amorphous silicon. *Journal of Non-Crystalline Solids* 299–302: 741–745.

Gerberich, W. W., et al. 2003. Superhard silicon nanospheres. *Journal of the Mechanics and Physics of Solids* 51(6): 979–992.

Grierson, D. S., E. E. Flater, and R. W. Carpick. 2005. Accounting for the JKR-DMT transition in adhesion and friction measurements with atomic force microscopy. *Journal of Adhesion Science and Technology* 19(3): 291–311.

Houston, Michael R., Roya Maboudian, and Roger T. Howe. 1995. Ammonium fluoride anti-stiction treatments for polysilicon microstructures. *The 8th Annual Conference on Solid-State Sensors and Actuators, 1995, and Eurosensors IX*, 210–213. Stockholm, Sweden: IEEE.

Hruby, Jill. 2001. LIGA technologies and applications. *MRS Bulletin* 26(4): 337–340.

Huh, Yoon, et al. 2005. Control of carbon nanotube growth using cobalt nanoparticles as catalyst. *Applied Surface Science* 249(1–4): 145–150.

Jia-Yang, Juang, and D. B. Bogy. 2005. Nanotechnology advances and applications in information storage. *Microsystem Technologies* 11(8—10): 950–957.

Johnson, K. L. 1987. *Contact Mechanics.* New York: Cambridge University Press.

Johnson, K. L., K. Kendall, and A. D. Roberts. 1971. Surface energy and the contact of elastic solids. *Proceedings of the Royal Society of London, Series A (Mathematical and Physical Sciences)* 324(1558): 301–313.

Klein, J., et al. 2004. Aluminium-induced crystallisation of amorphous silicon: Influence of the aluminium layer on the process. *Thin Solid Films* 451–452: 481–484.

Komvopoulos, K. 2003. Adhesion and friction forces in microelectromechanical systems: Mechanisms, measurement, surface modification techniques, and adhesion theory. *Journal of Adhesion Science and Technology* 17(4): 477–517.

Komvopoulos, K., N. Saka, and N. P. Suh. 1985. The mechanism of friction in boundary lubrication. *Transactions of the ASME Journal of Tribology Technology* 107(4): 452–461.

Komvopoulos, K., N. Saka, and N. P. Suh. 1986. Plowing friction in dry and lubricated metal sliding. *Journal of Tribology, ASME Trans.* 108(3): 301–303.

Kono, Y., et al. 2005. Study on nano imprint lithography by the pre-exposure process (PEP). *Proceedings of SPIE* 5753: 912–925.

Li, A. P., et al. 1998. Hexagonal pore arrays with a 50–420 nm interpore distance formed by self-organization in anodic alumina. *Journal of Applied Physics* 84(11): 6023–6026.

Liang, Jianyu, Hope Chik, Aijun Yin, and Jimmy Xu. 2002. Two-dimensional lateral super-lattices of nanostructures: Nonlithographic formation by anodic membrane template. *Journal of Applied Physics* 91(4): 2544–2544.

Liu, Jia J. 1997. Optimization of laser texture for the head-disk interface. *IEEE Transactions on Magnetics* 33(5): 3202–3204.

Loo, Yueh-Lin, Robert L. Willett, Kirk W. Baldwin, and John A. Rogers. 2002. Additive, nanoscale patterning of metal films with a stamp and a surface chemistry mediated transfer process: Applications in plastic electronics. *Applied Physics Letters* 81(3): 562–564.

Maboudian, R., and R. T. Howe. 1997. Stiction reduction processes for surface microma-chines. *Tribology Letters* 3(3): 215–221.

Maboudian, Roya, and Roger T. Howe. 1997. Critical review: Adhesion in surface micro-mechanical structures. *Journal of Vacuum Science & Technology B: Microelectronics Processing and Phenomena* 15(1), doi:10.1116/1.589247.

Martines, Elena, et al. 2005. Superhydrophobicity and superhydrophilicity of regular nanopatterns. *Nano Letters* 5(10): 2097–2103.

Masuda, H., and K. Fukuda. 1995. Ordered metal nanohole arrays made by a two-step replication of honeycomb structures of anodic alumina. *Science* 268(5216): 1466–1468.

Masuda, H., and M. Satoh. 1996. Fabrication of gold nanodot array using anodic porous alumina as an evaporation mask. *Japanese Journal of Applied Physics, Part 2 (Letters)* 35(1): 126–129.

Masuda, Hideki, Kenji Yasui, and Kazuyuki Nishio. 2000. Fabrication of ordered arrays of multiple nanodots using anodic porous alumina as an evaporation mask. *Advanced Materials* 12(14): 1031–1033.

Maugis, Daniel. 1992. Adhesion of spheres: The JKR-DMT transition using a Dugdale model. *Journal of Colloid and Interface Science* 150(1): 243–269.

Ming, W., D. Wu, R. van Benthem, and G. de With. 2005. Superhydrophobic films from raspberry-like particles. *Nano Letters* 5(11): 2298–1301.

Mirji, S. A. 2006. Octadecyltrichlorosilane adsorption kinetics on Si(100)/SiO2 surface: Contact angle, AFM, FTIR and XPS analysis. *Surface and Interface Analysis* 38(3): 158–165.

Murillo, R., et al. 2005. Fabrication of patterned magnetic nanodots by laser interference lithography. *Microelectronic Engineering* 78–79: 260–265.

Nair, R. P., et al. 2008. Wetting and adhesion properties of organic monolayer modified nano-topography-engineered surfaces. *Proceedings of the ASME/STLE 2007 International Joint Tribology Conference*, 1021–1023. New York: American Society of Mechanical Engineers.

Nair, R. P., and Min Zou. 2008. Surface-nano-texturing by aluminum-induced crystallization of amorphous silicon. *Surface & Coatings Technology* 203(5–7): 675–679.

Nast, O., and A. J. Hartmann. 2000. Influence of interface and Al structure on layer exchange during aluminum-induced crystallization of amorphous silicon. *Journal of Applied Physics* 88(2): 716–724.

Persson, B. N. J., and E. Tosatt, eds. 1996. *Physics of Sliding Friction.* Dordrecht: Kluwer Academic Publishers.

Peters, K. F., J. B. Cohen, and Yip-Wah Chung. 1998. Melting of Pb nanocrystals. *Physical Review B (Condensed Matter)* 57(21): 13430–13438.

Rabinowicz, E. 1995. *Friction and Wear of Materials.* New York: Wiley.

Raeymaekers, Bart, Izhak Etsion, and Frank E. Talke. 2007. Enhancing tribological performance of the magnetic tape/guide interface by laser surface texturing. *Tribology Letters* 27(1): 89–95.

Schmidt, M., et al. 1997. Experimental determination of the melting point and heat capacity for a free cluster of 139 sodium atoms. *Physical Review Letters* 79(1): 99–102.

Singh, D. P., A. K. Singh, and O. N. Srivastava. 2003. Formation and size dependence of germanium nanoparticles at different helium pressures. *Journal of Nanoscience and Nanotechnology* 3(6): 545–548.

Singh, R. A., et al. 2009. Bio-inspired dual surface modification to improve tribological properties at small-scale. *Applied Surface Science* 255(9): 4821–4828.

Song, Y., R. Premachandran Nair, M. Zou, and Y. A. Wang. 2009. Superhydrophobic surfaces produced by applying self-assembled monolayer on silicon micro/nano-textured surfaces. *Nano Research* 2: 143–145.

Song, Y., R. Premachandran Nair, M. Zou, and Y. A. Wang. 2010. Adhesion and friction properties of micro/nano-engineered superhydrophobic/hydrophobic surfaces. *Thin Solid Films* 518(4): 3801–3807.

Suh, N. P., and H. Sin. 1981. The genesis of friction. *Wear* 69(1): 91–114.

Tae-Sik Yoon, et al. 2004. Single and multiple-step dip-coating of colloidal maghemite (-Fe2O3) nanoparticles onto Si, Si3N4, and SiO2 substrates. *Advanced Functional Materials* 14(11): 1062–1068.

Tambe, N. S., and B. Bhushan. 2005. Nanotribological characterization of self-assembled monolayers deposited on silicon and aluminium substrates. *Nanotechnology* 16(9): 1549–1558.

Unruh, K. M., T. E. Huber, and C. A. Huber. 1993. Melting and freezing behavior of indium metal in porous glasses. *Physical Review B (Condensed Matter)* 48(12): 9021–9027.

Wang, Hengyu, et al. 2007. Friction study of a Ni nanodot-patterned surface. *Tribology Letters* 28(2): 183–189.

Wenzel, R. N. 1936. Resistance of solid surfaces to wetting by water. *Industrial and Engineering Chemistry* 28: 988–994.

Willis, Eric. 1985. Surface finish in relation to cylinder liners. *Wear* 109(1–4): 351–366.

Yao, Jimin, et al. 2004. Patterning colloidal crystals by lift-up soft lithography. *Advanced Materials* 16(1): 81–84.

Yee, Youngjoo, Kukjin Chun, and Duk L. Jong. 1995. *Polysilicon surface modification technique to reduce sticking of microstructures. Sensors and Actuators A: Physical* 52(1–3): 145–150.

Yoon, Eui-Sung, Seung H. Yang, Hung-Gu Han, and Hosung Kong. 2003. An experimental study on the adhesion at a nano-contact. *Wear* 254(10): 974–980.

Zhu, Lingbo, et al. 2006. Optimizing geometrical design of superhydrophobic surfaces for prevention of Microelectromechanical System (MEMS) stiction. *Proceedings of the Electronic Components and Electrical Conference*, doi: 10.1109/ECTC.2006.1645795

Zou, M., et al. 2005. Adhesion and friction studies of a selectively micro/nano-textured surface produced by UV assisted crystallization of amorphous silicon. *Tribology Letters* 20(1): 43–52.

Zou, M., L. Cai, and H. Wang. 2006. Adhesion and friction studies of a nano-textured surface produced by spin coating of colloidal silica nanoparticle solution. *Tribology Letters* 21(1): 25–30.

Zou, M., L. Cai, H. Wang, D. Yang, and T. Wyrobek. 2005. Adhesion and friction studies of a selectively micro/nano-textured surface produced by UV assisted crystallization of amorphous silicon. *Tribology Letters* 20: 43–52

Zou, M., W. Seale, and H. Wang. 2005. Comparison of tribological performances of nano- and micro-textured surfaces. *Proceedings of the Institution of Mechanical Engineers, Part N (Journal of Nanoengineering and Nanosystems)* 219(3): 103–110.

Zou, Min, et al. 2006. Ni nanodot-patterned surfaces for adhesion and friction reduction. *Tribology Letters* 24(2): 137–142.

7 Environmental Effects in Tribology

Seong H. Kim

CONTENTS

7.1 INTRODUCTION

The chemistry involved in tribological interactions of two rubbing solids is complicated [1,2]. Mechanical contact, deformation, and shear at the interface create dynamic and often chemically reactive conditions for the atoms at the surface and within the subsurface [1]. Unless tribological contacts are made in ultra-high vacuum (UHV) conditions with no molecules adsorbed on the surface, or in ultra-high purity gases with absolutely no reactive impurities, there always will be third "chemical" components involved in the tribology in addition to the two sliding and counter-sliding solid surfaces. These third components can drastically alter the tribological responses of the solid interfaces. Those are the molecules impinging onto the solid surface from the surrounding space. If a liquid lubricant is utilized to mitigate the friction and wear problems between two solid contacts, then it is obvious that the tribological interface consists of three components; that is, solid--liquid--solid. Even if there is no liquid lubricant added to the system, there are always molecules adsorbed onto the solid surface from the gas phase that act as a third tribological component.

Glass windows provide an everyday example of complex surfaces. If one looks outside through a hazy glass window, one would easily notice stains on the window. After cleaning the window thoroughly, the window would look clean. However, even the clear window would only be optically clean, not chemically clean. Most glass windows are made from soda-lime glass, which is an amorphous network of silicon oxide (about three quarters) and other minerals such as sodium, calcium, and so on. The surface energy of the clean oxide is so high that the surface will be covered instantaneously by a water layer in a humid ambient [3]. The gas uptake by the solid surface is called adsorption. If one touches the glass surface with a finger, the adsorbed water layer will make contact with the finger first, before the solid surface does (although we would not feel the difference). In addition, the chemical composition of the soda-lime glass can vary depending on cleaning methods [4]. Water can cause sodium to leach from the subsurface region of glass, resulting in changes in the pH of water near the glass surface as well as changes in the mechanical and chemical properties of the glass surface [5,6]. The adsorbed water on glass cannot be completely removed unless the glass is heated to a high temperature in vacuum [7]. As soon as the glass is retrieved from the vacuum chamber, the high surface energy nature of glass will allow the immediate uptake of molecules from the environment (such as water and organic contaminant from ambient air). This example illustrates that tribological properties of a specific surface may not be explained or understood using only the chemical composition and mechanical properties of the bulk phase.

Gas adsorption is a well-established field in chemistry and chemical engineering and is the core subject of heterogeneous catalysis, separation, and corrosion processes [8,9]. Not surprisingly, examples of gas adsorption effects on adhesion, friction, and wear behaviors of solid materials (which are the three main measurables in tribology) are frequently encountered in the literature. However, the level of understanding or control of these effects is far inferior in tribology compared to pure chemical science and engineering fields. In purely chemical systems, the adsorption and reaction of gas molecules on solid surfaces are governed by thermodynamics (i.e., pressure and temperature of the system as well as the surface chemistry and structure) [8,10]. In some cases, non-thermal processes involving photons or electrons drive chemical reactions, but their mechanisms are relatively well understood through molecular orbital theories [11]. In tribology, however, the situation is much more complicated due to the presence of mechanical stress and/or shear applied locally to the solid surface [12,13]. Stress and shear can alter the chemical equilibrium or energetics of molecular orbitals involved in chemical reactions. The mechanical force applied to the interface can alter the thermodynamic equilibrium of the adsorbed molecules or be channeled into the reaction coordinate of the adsorbed molecules. Molecular-level understanding of these mechanically driven chemical reactions is minuscule.

In order to set the foundation for a deeper understanding of chemical effects at tribological interfaces, it is important to review environmental effects reported in the current tribology literature. It is also important to discuss the previously accepted explanations for the environmental effects, and how to further improve and advance tribochemistry knowledge. The term "tribochemistry" is often used to collectively mean chemical reactions involved in or taking place at tribological interfaces. The lack of tribochemical understanding in the past was in part due to ignorance of adsorbed molecules at the interface. This ignorance was the case until surface analytical tools became readily available and were applied to tribological studies [14]. Nowadays, surface analysis techniques such as Auger electron spectroscopy (AES), x-ray photoelectron spectroscopy (XPS), time-of-flight secondary ion mass spectrometry (ToF-SIMS), and so on are readily employed in tribology research [15–18]. Although these techniques are helpful, they are still not sufficient to bring tribochemical understanding to the molecular level. State-of-the-art surface science techniques provide a means to control or identify the chemical nature of the solid interface before and after tribological contacts are made. However, these methods often fail to work at the asperity contacts made during tribological actions since the interfaces are buried and not readily accessible to spectroscopic probes such as photons and electrons. Some techniques can be applied directly during sliding or contact [19], but they often lack the sub-monolayer detection sensitivity or spatial resolution in chemical differentiation. Another reason for the lack of tribochemical understanding could be the improper use of surface analysis results in interpreting tribological data, or finding correlations between the tribological observations and chemical aspects of interfaces due to bias or conventional expectations.

This review consists of three parts. The first part surveys some of the environmental effects reported for various tribological interfaces involving metals, oxides, ceramics, carbon coatings, and solid lubricants. It is impossible to cover all of the tribochemistry examples found in the literature, so only selected examples are included that would help the reader follow the current status and evolution of our understanding of environmental effects on tribology. The second part is a brief review of this author's own work on a specific system—the effects of alcohol and water vapors on capillary adhesion, friction, and wear of native silicon oxide surfaces. In this section, the focus is to demonstrate how surface analysis can be helpful and used to unveil fundamental insights of environmental effects observed for silicon oxide surfaces in alcohol and water vapor environments. The third section looks into a few examples where the environmental effects can be used to mitigate friction and wear of solid surfaces.

It should be recognized that the tribological properties (adhesion, friction, and wear) are not intrinsic material properties. These properties can and will vary, depending on the sample preparation and measurement conditions, exemplified by the previously mentioned glass window. Consequently, care must be taken when comparing the experimental data from different groups. Another important aspect to be noted is that the chemical properties of solid surfaces are not size-dependent properties. They are, rather, scale-invariant properties—that is, they do not depend on the system size or the amount. However, the manifestation of the effects of interfacial chemistry in tribological measurements can vary with the size of the contacts (nano- and micro-scale versus macro-scale) and the magnitude of applied contact load. Therefore, one must be cautious when the insight obtained from one system is to be generalized or applied to other systems.

7.2 SURVEY OF ENVIRONMENTAL EFFECTS IN TRIBOLOGY

7.2.1 METALS

Clean metal surfaces have very high surface energies [14], so they readily uptake gases from the environment. This lowers the free energy of the system. Quantitative investigations on the effects of adsorbed gas on friction and wear of metal surfaces were pioneered by Buckley at NASA in the 1960's [20]. He studied the effects of clean and adsorbate-covered tungsten surfaces in UHV conditions where the nature and amount of the adsorbed species can be controlled precisely [21]. These tests showed that the chemisorption (chemical adsorption) of simple gases such as hydrogen, oxygen, carbon dioxide, and hydrogen sulfide on the tungsten surface can significantly reduce the friction of tungsten single-crystal surfaces (Table 7.1). Oxygen provides the largest reduction and hydrogen provides the least. Also, the structure of tungsten impacts the effectiveness of the gas on friction reduction: the open surface structure experiences a larger reduction than the closer-packed surface structure. In the homologous series of hydrocarbons from methane through decane, the friction coefficient of tungsten decreases with the increasing number of carbon atoms (Figure 7.1). The carbon–carbon bond order of the hydrocarbon adsorbate also influences the friction, with the order of decreasing friction being H_3CCH_3, $H_2C=CH_2$, then HCCH (Table 7.1). Detection of desorbing species during sliding with a mass spectrometer indicated hydrogenation of unsaturated hydrocarbons. It was also demonstrated that the presence of multiple physisorbed (physically adsorbed) layers beyond a simple chemisorbed monolayer can reduce the friction markedly. At atmospheric pressure, metal oxides and sulfides can form as a result of heat generated by sliding surfaces, preventing stick-slip motion and giving lower friction than does the chemisorbed layer in UHV conditions. Later, Buckley combined the UHV

TABLE 7.1

Influence of Various Chemisorbed Gases on Friction Coefficient of Tungsten in Vaccum.[21]

		Friction Coefficient		
Chemisorbed Gas		**W(110) Surface**	**W(210) Surface**	**W(100) Surface**
Simple Gases	None	1.33	1.90	3.00
	H_2	1.25	1.33	1.66
	O_2	0.95	1.00	1.30
	CO_2	1.15	1.15	1.40
	H_2S	1.00	—	1.35
C_2 Hydrocarbons	H_3CCH_3	1.10	1.10	1.25
	$H_2C=CH_2$	0.88	0.85	1.20
	HCCH	0.70	0.66	1.00

Note: Counterface = W(100) surface; load = 50 g; sliding velocity = 0.001 cm/s; temperature = 20 °C; ambient pressure = 10^{-10} Torr.

Source: Buckley, D. H., *J. Appl. Phys.* 39 (1968): 4224–4233.

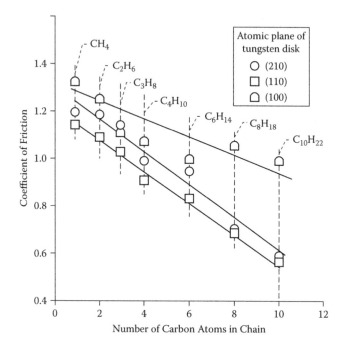

FIGURE 7.1 Coefficient of friction for single-crystal tungsten (100) surface sliding on single-crystal tungsten disks with (100), (110), and (210) orientations with chemisorbed monolayers of hydrocarbons. Load = 50g, sliding velocity = 0.001 cm/sec, temperature =20°C, ambient pressure = 10^{-10} Torr. (From Buckley, D. H. 1968. *J. Appl. Phys.* 39: 4224–4233.)

tribometer with AES for chemical analysis and quantification of the tribo-tested surface [22]. This instrument was used to show that the static friction coefficient of metals decreases with increasing adsorbate concentration, and that the static friction coefficient is independent of the type of metal and the adsorbate species for equal concentrations of adsorbate [23]. These pioneering works carried out at NASA set the foundation for modern tribochemistry study.

Following these momentous works, attempts were made to establish generalized chemical theories that can explain environmental effects observed for various metals tested in different ambient conditions. For example, some correlations were observed between atmospheric gas chemisorption and wear of metals (Figure 7.2) that were different from the monotonic decrease or insignificant change of the friction coefficient as the atmospheric gas pressure increases [24]. Interestingly, the wear–atmosphere curve had a maximum point at a pressure between 10^{-5} Pa (~10^{-10} atm) and 10^5 Pa (~1 atm) when the gas molecules were strongly chemisorbed onto the metal surfaces. This trend was explained with the competition between particle adhesion and chemical bonding strength. In this argument, the adhesion force between the metal surfaces is strong in a high vacuum and the wear particles cannot easily be ejected from the sliding surface. Thus, the overall wear volume is low but the friction coefficient is high. With increasing atmospheric pressure, the chemisorbed gas molecules begin to inhibit the adhesion of metals, allowing the particles to be easily

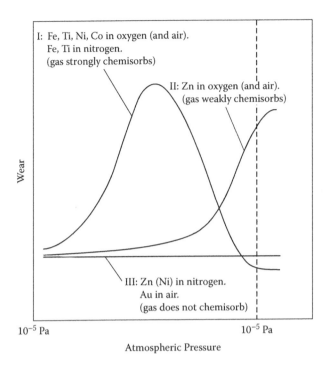

I: Fe, Ti, Ni, Co in oxygen (and air).
Fe, Ti in nitrogen.
(gas strongly chemisorbs)

II: Zn in oxygen (and air).
(gas weakly chemisorbs)

III: Zn (Ni) in nitrogen.
Au in air.
(gas does not chemisorb)

Wear

10^{-5} Pa 10^{-5} Pa

Atmospheric Pressure

FIGURE 7.2 Three types of atmospheric characteristics in adhesive wear of metals. (From Mishina, H. 1992. *Wear* 152: 99–110.)

removed from the sliding surface. This could result in an increase of wear volume with increasing atmospheric pressure even though the chemisorbed gas molecules reduce the friction coefficient. Upon further increase of the atmospheric gas pressure, a stable oxide layer is formed, which is resistant to wear. This is a phenomenological explanation proposed to justify the observed experimental data. It does not provide any molecular-level understanding of the correlation between wear and oxygen chemisorption.

The effect of relative humidity (RH) on metal wear is also complicated. Goto and Buckley reported the fretting wear volume of pure metals as a function of RH [25]. Here, the fretting refers to minute relative motion of surfaces caused by repetitive vibrations whose amplitude is much smaller than the asperity contact size. Most metals tested, except aluminum and nickel, exhibited maxima in low relative humidity regions (Figure 7.3). The authors assumed that in dry conditions the metal surface follows the oxidative wear mechanism, and when a proper amount of water vapor adsorbs onto the metal surface, oxygen adsorption is reduced and adhesive wear of parent metal-to-metal contact can readily occur, resulting in relatively larger wear particles. As the RH in air increases further, a water film forms on the surface. Then, the water and oxide films lubricate and mitigate the friction, resulting in less fretting wear in the high RH region. Again, this is a plausible speculation, but there is no molecular or atomic level evidence to support the proposed details or guarantee the extension of the justification to other systems.

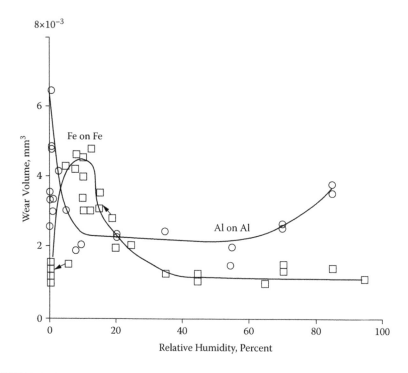

FIGURE 7.3 Fretting wear volume as a function of relative humidity. Contact load = 2.94N, fretting amplitude = 80 μm, frequency = 60Hz, number of fretting cycles = 5.04 × 105. (From Goto, H., and Buckley, D. H. 1991. *Wear* 143: 15–28.)

Recent tribochemical studies on environmental effects of the friction and wear of metal surfaces focus on rather specific systems related to certain engineering applications, stricter controls of experimental variables, and use of spectroscopic techniques that can provide chemical information as well as elemental concentration of the adsorbate species (such as XPS) [26]. The surface coverage dependence of friction was revisited, and it was found that the saturated monolayers of adsorbed atoms cannot prevent direct metal–metal contact, which dictates the subsequent sliding behavior of the interface (Figure 7.4) [27]. In contrast, the adsorbed molecular layers can act as good lubricants (Figure 7.4) [28]. An interesting engineering example is the tribology of metal alloys in carbon dioxide environments. This may play an important role in NASA's missions to explore the Martian surface, since the CO_2 pressure in the atmosphere of Mars is 10~15 Torr, which is 30–50 times higher than the partial pressure of CO_2 in the Earth's atmosphere [29]. CO_2 appears to effectively lubricate 52100/440C bearing steels [30]. XPS analysis revealed that these steel surfaces lubricated with CO_2 contain iron carbonate and/or bicarbonate, which is believed to be responsible for the reduction of friction and wear (Figure 7.5) [30]. There seems to be an optimum pressure, between 0.2 and 0.5 atm, for reducing friction and wear effectively. It is understandable that low CO_2 pressure may not be sufficient to produce the carbonate and/or bicarbonate species at the surface. The increased pressures of CO_2 may catalyze chemical wear processes.

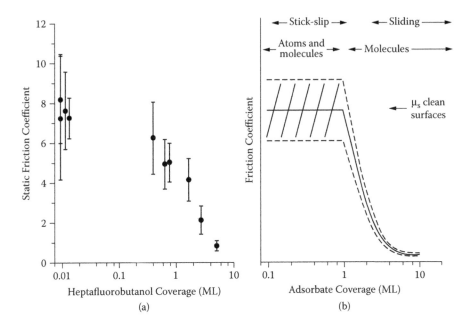

FIGURE 7.4 (a) Plot of friction coefficient versus heptofluorobutanol coverage. Temperature = 100K, applied load = 25–40 mN, slide rate = 20 µm/sec. (b) Schematic diagram illustrating the dependence of the friction coefficient between Cu(111) surfaces on adsorbate coverage. (From McFadden, C. F., and Gellman, A. J. 1998. *Surf. Sci.* 409: 171–182.)

Some metals are more reactive to certain gases than others. Molybdenum is highly reactive with oxygen, sulfur dioxide, and hydrogen sulfide [31]. Reaction layers on Mo are mostly one- or two-layer-thick MoO_2 and/or MoS_2 species. Upon sliding, much thicker tribochemical films with the same elemental compositions are produced on the Mo surface, which can mitigate friction and wear of Mo. Similar effects are also observed for iron in alkyl halide gas environments and steels coated with tricresylphosphate (TCP) [32–35]. These reactions are relevant to extreme-pressure and anti-wear additives used along with liquid lubricants.

Noble metals—that is, the metals of groups VIII and IB of the second and third transition series of the periodic table—have outstanding resistance to oxidation and other chemical reactions. One may expect no environmental effects for metals in these groups. It should be noted, however, that the noble metals are key elements in mainstream catalytic conversion processes in chemical industries [36–37]. Although they do not easily form chemical compounds, they can facilitate other chemical reactions, such as hydrogenation, dehydrogenation, hydrogenolysis, and oxidation reactions of hydrocarbons. Therefore, it would be a mistake to assume that noble metals are chemically inert in tribological conditions.

Reactions under tribological conditions appear to be different from thermal catalytic reactions. Rubbing palladium against aluminum oxide in an ethylene/oxygen gas mixture can oxidize ethylene to carbon dioxide and water even at room temperature [38]. In contrast, heating a palladium wire to >600 K in a similar environment

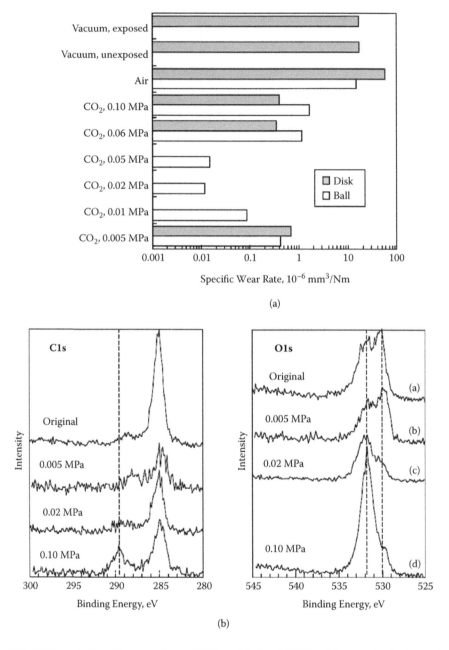

FIGURE 7.5 (a) Specific wear rates of 52100 steel disk and 440C stainless steel ball obtained under different CO_2-containing environments. (b) C 1s and O 1s XPS spectra of the original surface and the worn surfaces of disk under 0.005 MPa, 0.02 MPa, and 0.10 MPa of CO_2. (From Wu, X., Cong, P., Nanao, H., Minami, I., and Mori, S. 2004. *Tribol. Lett.* 17: 925–930.)

produces carbon monoxide and hydrogen. These results indicate that the ethylene is oxidized on palladium via tribochemical pathways that are different from the normal catalytic pathways. Electrical contacts can also be influenced by tribological conditions. The electrical resistance of palladium fretting contact increases in dry air due to oxidation, but the increase of contact resistance can be suppressed by introducing hydrocarbons into the environment [39]. Again, this result indicates that the hydrocarbons are tribochemically reacting at the nascent palladium surface produced by fretting wear. Even diamond can wear when rubbed against certain transition metal surfaces, including noble metals [40]. The chemical wear of diamond surfaces requires the dissociation of the C–C bond, which could occur via friction- and wear-induced hydrogenolysis on catalytically active metal surfaces generated by sliding against a diamond surface.

Gold is the most inert metal in normal environments, but even gold is not inert in tribological environments. When a gold surface is rubbed with alumina in the presence of cyclohexene gas, the nascent gold surface exposed by the sliding hydrogenates cyclohexene to cyclohexane (Figure 7.6a) [41,42]. In contrast, the silver-palladium alloy surface dehydrogenates cyclohexene to benzene in the same frictional conditions (Figure 7.6b) [41,42]. Without sliding at the interface, cyclohexene will not react on clean gold surfaces.

7.2.2 CERAMICS

Ceramics are mechanically harder, thermally more stable, and chemically less reactive than metals. These properties make ceramics very attractive in tribological applications. However, they still are not free from tribochemical reactions involving molecules from the surrounding medium or environment [2].

In dry argon environments, silicon nitride undergoes purely mechanical wear that consists of microfractures at the surface, leaving a rough wear track. When sliding is made in humid argon, the wear track is smooth and the surface is covered by amorphous silicon oxide [43]. At high temperatures (>650°C), the effect of humidity is not observed because water vapor does not adsorb onto the surface [44]. When sliding in liquid water, the wear track becomes so smooth that hydrodynamic lubrication by a water film occurs, providing an ultra-low friction coefficient (Figure 7.7) [45,46]. These processes are explained by the oxidation of silicon nitride by reaction with water ($Si_3N_4 + 6 H_2O \rightarrow 3 SiO_2 + 4 HNO_3$) and the dissolution of silica in the form of silicic acid ($SiO_2 + 2 H_2O \rightarrow Si(OH)_4$) [47,48]. These reactions preferentially occur at the asperity contacts, so the asperities are removed as the tribochemical wear progresses in the contact area.

Oxide ceramics, such as alumina and zirconia, do not undergo oxidation reactions that are responsible for the decrease in wear of silicon nitride, but they are still susceptible to environmental tribochemistry. Tetragonal zirconium oxide, stabilized by yttria doping, has a high fracture toughness, making it much more resistant to mechanical wear than silicon nitride. Thus, its wear rate in dry conditions is several orders of magnitude lower. Depending on the experimental test conditions, the wear of yttria-stabilized zirconia (YSZ) can either increase or decrease. Sliding in humid air and in water can induce hydroxide formation at the intergranular interfaces, which

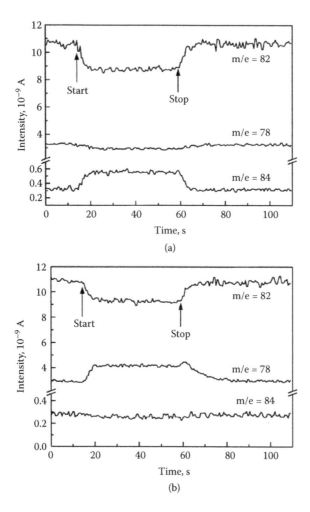

FIGURE 7.6 Variation of the mass spectrometer signal when (a) a gold disk and (b) a Ag-Pd alloy disk were rubbed by alumina balls in cyclohexene atmosphere. Gas pressure = $8 \times 10{-}4$ Pa, ball diameter = 6.35 mm, load = 4 N, sliding velocity = 8cm/sec, m/e 82 = cyclohexene, m/e 78 = benzene, m/e 84 = cyclohexane. (From Wu, X., Kobayashi, N., Nanao, H., and Mori, S. 2005. *Tribol. Lett.* 18: 239–244.)

can accelerate the microfracture wear process [49]. In contrast, the same hydroxide formation was reported to be responsible for reduction in friction and wear of tough zirconia in a fretting test [50]. The wear behavior of alumina is also reported to be sensitive to humidity due to its susceptibility to stress corrosion cracking [51,52].

Similar to water, simple organic molecules alter the friction and wear behavior of ceramics. The friction coefficient and wear rate of silicon nitride and silicon carbide decrease significantly when sliding occurs in liquid alcohols compared to sliding in water [53]. The lubrication effect of alcohol is augmented as the alkyl chain length increases. The alcohol lubrication effect can also be observed in UHV conditions,

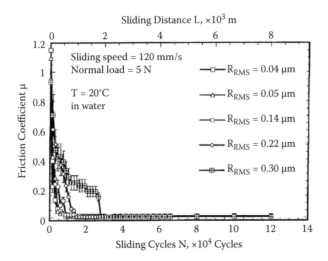

FIGURE 7.7 Variation of friction coefficient with sliding cycles of Si3N4 with different initial roughness under water. (From M. Chen, K. Kato, and K. Adachi, 2001. *Wear* 250: 246–255.)

where there is no liquid or vapor alcohol and only adsorbed alcohol molecules exist on the surface. In a spectroscopic and atomic-resolution imaging study that was coupled with friction measurements of ethanol adsorbed on vanadium carbide, it was found that the chemisorbed monolayer of ethanol is not sufficient for reduction of friction, and the decrease of friction coincides with the formation of an amorphous deposit on the surface at high ethanol doses (Figure 7.8) [54].

The advance of technology requires full understanding and control of tribological behavior in a wide range of environmental conditions. Fluorohydrocarbons are now used as a refrigerant instead of chlorofluorocarbons due to environmental issues, so the effects of these fluorinated molecules on friction and wear are of interest. When a silicon carbide disk was slid against an alumina ball in CF_3CH_2F (HFC-134a) gas, friction and wear decreased dramatically as the gas pressure was increased from vacuum to one atmosphere, whereas the amount of carbon-containing compounds and fluorides on the wear track rapidly increased [55]. The tribochemical reactions produced metal fluorides and organic compounds, which were correlated with the friction reduction and wear mitigation. Again, the nascent oxide surface produced by friction and wear appears more reactive than the normal, clean surface prepared by chemical methods [56].

Even simple hydrocarbon gases can influence the tribological behavior of ceramic surfaces. Carbon deposits can be formed on tribo-contacts of silicon nitride surfaces in ethylene or other carbonaceous feed gases at temperatures near 500°C, and they can effectively lubricate the sliding interface (Figure 7.9) [57]. It was reported that the friction and wear of alumina became minimal at a certain pressure range of n-butane. At this condition, polymer-like deposits were formed in the wear track, and the activity of the energetic particle emission was a maximum, implying that these two processes were related [58].

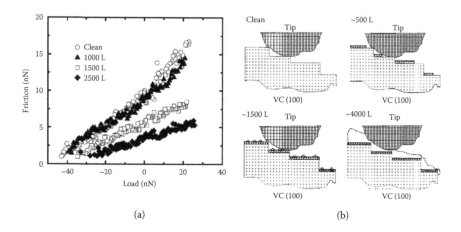

(a) (b)

FIGURE 7.8 (a) Friction load plots measured with AFM under UHV conditions between a silicon nitride probe tip and the VC(100) surface as a function of ethanol exposure (L = Torr × sec). (b) Schematic model of the formation of the boundary layer film produced by the room temperature reaction of ethanol with the VC(100) surface. At exposure greater than ~1000L, a significant reduction in friction is observed and correlates with the formation of the first monolayer of the boundary layer film. (From Kim, B. I., Lee, S., Guenard, R., Torres, L. C. F., Perry, S. S., Frantz, P., and Didziulis, S. V. 2001. *Surf. Sci.* 481: 185–197.)

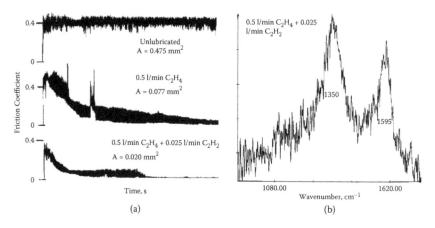

(a) (b)

FIGURE 7.9 (a) Friction traces for silicon nitride pin-on-disk contacts in different hydrocarbon environments. (b) Raman spectrum of the carbon deposit on the disk. (From Lauer, J. L., Blanchet, T. A., Vleck, B. L., and Sargent, B. 1993. *Surf. Coat. Technol.* 62: 399–405.)

7.2.3 CARBON SURFACES

Graphite had initially been considered to be an effective dry lubricant. Its layer structure was generally believed to give low shear stress, since the bonding between the basal planes is weak. It was later realized that the lubrication effect of graphite is observed only in certain environments containing, for example, oxygen or water [59]. The slipperiness of graphite is not an intrinsic property; it depends on gases and vapors in the environment that can interact or react with carbon atoms (Figure 7.10) [60]. It is now known that uptake of vapor molecules increases the basal plane spacing at the graphite surface and weakens the interlayer bonding, which makes interlayer shearing more feasible [61].

The extreme hardness of diamond makes it an ideal material for many applications, including, for example, cutting tools used in rock drilling and precision machining. However, diamond cannot be used for cutting steels and ferrous alloys, or even some soft transition metals, due to tribochemical reactions [40]. Like graphite, diamond exhibits good lubricity in humid conditions, but the friction is high in

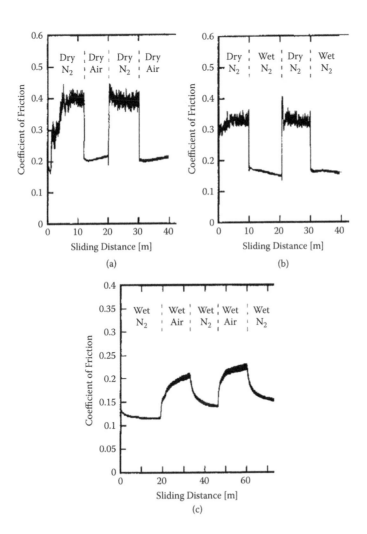

FIGURE 7.10 Change in the friction of graphite upon cycling in (a) dry nitrogen and dry air, (b) dry nitrogen and humid nitrogen, and (c) humid nitrogen and humid air environments. (From Yen, B. K. 1996. *Wear* 192: 208–215.)

vacuum [62–65]. Unless the dangling bond at the diamond surface is passivated effectively, diamond will give a high friction coefficient.

Amorphous carbon (a-C), also known as diamond-like carbon (DLC), is widely used in magnetic hard disk industries. Its chemical composition and structure can vary widely, depending on the deposition methods and precursors used. In general, it consists of a mixture of sp^2 and sp^3 hybridized carbons with varying amounts of hydrogen [66]. Knowing the environmental sensitivity of graphite and diamond, it is not surprising that the friction and wear properties of a-C and DLC can vary with testing environments and preparation methods [67,68]. Hydrogen-rich DLC can give a friction coefficient as low as 0.003 in UHV, but its friction coefficient increases to ~0.1 as the water and oxygen partial pressures increase to the ambient air level [69]. In contrast, hydrogen-poor DLC gives a friction coefficient close to unity in UHV, but its friction decreases to ~0.1 in ambient air conditions (Figure 7.11) [69]. The friction coefficient of hydrogen-rich DLC films is highly dependent on the gas exposure time between consecutive sliding cycles, indicating that it is determined by the balance between gas adsorption and removal kinetics [70,71]. These environmental effects must be related to the fact that the DLC surface gets oxidized or hydroxylated

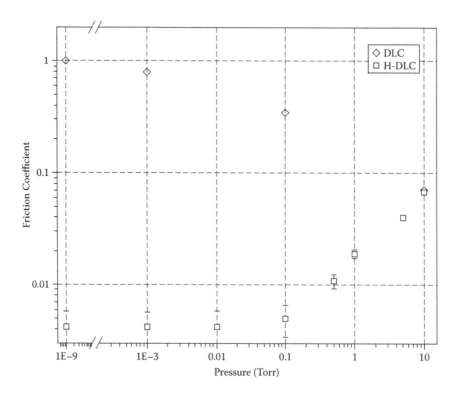

FIGURE 7.11 Effect of water on two DLC films with different hydrogen contents. Although their intrinsic friction coefficients are distinctly different in vacuum, the values converge with increasing H_2O pressure. (From Kim, H. I., Lince, J. R., Erylmaz, O. L., and Erdemir, A. 2006. *Tribol. Lett.* 21: 53–58.)

as it is rubbed in oxygen or water [72,73]. The friction coefficient of a-C can be reduced to ~0.01 by introducing hydrogen into the environment [74,75]. The environmental sensitivity of a-C affects the durability of magnetic hard disks where a-C is used as a protective layer [76–78].

7.2.4 SOLID LUBRICANTS

Molybdenum disulfide is widely used as a solid lubricant. It is similar to graphite in that it has a layer structure, but it exhibits the opposite environmental effects. In UHV or inert conditions, molybdenum disulfide can easily be transferred to the counterface, and the friction coefficient of two molybdenum-disulfide coated surfaces can be less than 0.01. However, the introduction of atmospheric air increases the friction coefficient by more than a factor of 10 [79].

One could come up with an innovative strategy to make a solid lubricant film capable of working in a wide range of environmental conditions—in dry nitrogen, in humid air, or at high temperature. This has been demonstrated with a composite coating consisting of yttria-stabilized zirconia in a gold matrix with encapsulated nano-sized reservoirs of MoS_2 and DLC [80]. These composite structures allowed the coating to adapt its chemistry to the lowest frictional composition in response to the changes of the tribo-test conditions. After sliding in humid air, the surface of the composite coating was dominated by graphitic carbon, which provides a friction coefficient of 0.10–0.15. After friction tests in dry nitrogen, the surface was covered with hexagonal MoS_2 with a friction coefficient of 0.02–0.05. When slid in air at 500°C, metallic Au covered the coating surface and gave a friction coefficient of 0.10–0.20 [80]. A composite coating made of MoS_2, Sb_2O_3, and carbon also showed a similar environmental adaptation capability [81].

As reviewed here, environmental effects on tribological systems are ubiquitous. Metals, alloys, ceramics, and carbon-based materials are all subject to tribochemical effects when tested or operated in various environments. The examples discussed in this section are only a small fraction of the environmental effects reported in the literature. They clearly reveal that the tribological properties of a system are not an intrinsic material property; the properties depend on testing or operating conditions (i.e., applied load, sliding speed, temperature, and chemical composition of the surrounding atmosphere). Therefore, generalizations made from a specific system for a different material or test environment could lead to erroneous conclusions. With this alert in mind, we will now scrutinize in depth the fundamentals of the effects of alcohol and water vapor environments on tribological behavior of silicon oxide contacts.

7.3 EFFECTS OF ALCOHOL AND WATER VAPOR ON ADHESION, FRICTION, AND WEAR OF SILICON OXIDE

Chemical composition of the ambient gas can drastically influence adhesion, friction, and wear so that tribological data cannot be fully explained with solid surface properties alone (modulus, roughness, surface energy, etc.) unless the solid surface is completely inert from interactions with gas phase molecules. This section reviews a series of the author's work on the effects of alcohol and water vapors on tribological properties of silicon oxide surfaces and addresses them in a cohesive manner so that the reader can grasp the full spectrum of tribochemical issues involved in this specific system. The nano-scale fundamentals of capillary adhesion phenomena are explained in Section 7.3.1. Next, the lubrication effects under equilibrium adsorption conditions of alcohol vapors are described in Section 7.3.2. Finally, understanding the tribochemistry of silicon oxide surfaces in the ambient containing alcohol and water vapor is discussed in Section 7.3.3.

7.3.1 CAPILLARY ADHESION

(For a thorough background discussion of surface energy and forces, please refer to the chapter by Carpick in this volume.)

On fully hydroxylated silicon oxide surfaces exhibiting a water contact angle of ~0°, water or alcohol will mostly be physisorbed. These adsorbed molecules will have hydrogen bonding interactions with surface hydroxyls, and their molecular structure will be mostly intact. Even if the partial pressure of water and alcohol in the ambient gas is lower than its saturation pressure, a condensed phase can form in the narrow gap between two solid surfaces if the vapor molecule has a strong affinity toward the solid surface. This phenomenon is known as capillary condensation and is explained by the equilibrium relationship between the Laplace pressure caused by the curvature of the meniscus of the condensed phase and the partial pressure of the condensing molecule in the vapor phase.

The force exerted by the capillary meniscus around the asperity contact can be measured with atomic force microscopy (AFM). Figure 7.12a displays the capillary force measured with AFM for silicon oxide surfaces as a function of alcohol vapor pressure in a dry argon environment [82]. The measured force is normalized with $4\pi R\gamma$ where R is the local radius of the AFM tip and γ is the surface tension of the liquid alcohol. The experimental data clearly show that the capillary force strongly depends on the relative partial pressure of alcohol with respect to its saturation vapor pressure (p/p_{sat}). The capillary force decreases with increasing p/p_{sat} and with increasing alcohol chain length.

Now, let's take a look at how this behavior can be explained theoretically. The AFM tip is often modeled as a sphere since it is the simplest geometry to model mathematically and represents a sharp AFM tip end reasonably well [83]. The model system thus becomes a sphere-on-flat geometry (Figure 7.13a). There are two components contributing to the total capillary force (F_{cap}). One is due to the Laplace pressure (ΔP_L) across the meniscus surface, and the other is due to the surface tension (γ) along the meniscus surface. The force due to the Laplace pressure (F_{lp}) and the force due to the surface tension (F_{st}) can be calculated using the following equations, respectively:

$$F_{lp} = -\Delta P_L \text{ meniscus area} \tag{7.1}$$

$$F_{st} = \gamma \times \text{meniscus circumference} \times \sin\varphi \tag{7.2}$$

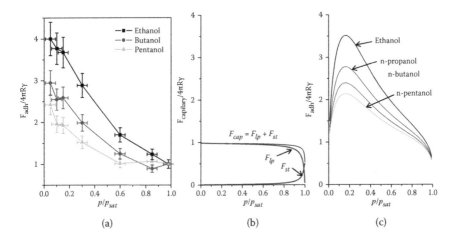

FIGURE 7.12 (a) Capillary adhesion versus p/psat for ethanol, n-butanol, and n-pentanol. All data points are normalized to the saturated condition where Fadh = 4πRγ. (From Asay, D. B., and Kim, S. H. 2007. *Langmuir* 23: 12174.). (b) Total capillary force as a function of relative partial pressure, as well as Laplace pressure and surface tension components, calculated for a sphere with R = 100 nm on a flat surface without taking into account the adsorption isotherm of ethanol. (c) Capillary forces as a function of relative partial pressure of ethanol, n-propanol, n-butanol, and n-pentanol calculated using the Laplace-Young exact solution method, taking into account the adsorption isotherm of each alcohol. (From Asay, D. B., de Boer, M. P., and Kim, S. H. 2010. *J. Adh. Sci. Technol.*, 24: 2363–2382).

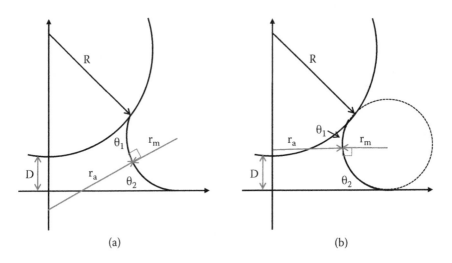

FIGURE 7.13 Typical geometry considered for a liquid bridge between a sphere and a flat substrate. (a) Exact geometry. (b) Circular approximation.

where φ is the angle between the line normal to the liquid meniscus surface and the surface normal direction along which the force is measured. The Laplace pressure is calculated from the Young-Laplace equation, which contains two principal radii of the meniscus:

$$\Delta P_L = \gamma \left(\frac{1}{r_a} + \frac{1}{r_m} \right) = \frac{\gamma}{r_e} \qquad (7.3)$$

where r_a (azimuthal radius) and r_m (meridional radius) are the principal radii of curvature of the meniscus surface and r_e is the effective radius of curvature of the meniscus. Since the liquid meniscus is in equilibrium with the vapor with a constant pressure, its surface is isobaric. This means that r_e is constant along the meniscus surface, while r_a and r_m vary along the surface.

One of the most commonly used equations to estimate or describe the capillary force is $F_{cup}=4\pi R\gamma\cos\theta$ where R, γ, and θ are the radius of the sphere, the liquid surface tension, and the contact angle of the liquid on the solid surface, respectively [84,85]. This equation is derived with several assumptions. The meniscus is assumed to be circular (Figure 7.13b), so only two fixed values are used to describe r_a and r_m. The r_m term is now the radius of the circle representing the external curvature of the meniscus, and the r_a term is the shortest distance from the meniscus center to the meniscus surface. If $r_a \gg r_m$, then $\Delta P_L \approx \gamma/r_m$ and the meniscus area is πr_a^2. If $R \gg r_a$, then $\pi r_a^2 \approx 2\pi R \cdot d$ (where d = height of the meniscus) and $d \approx 2r_m \cdot \cos\theta$ [84]. The F_{st} term is assumed to be negligible since it is usually much smaller than F_{lp} when p/p_{sat} is far below saturation. The main limitation of this approach in describing nanoscale capillary forces is the over-simplification with the relative size difference. Notice that this model has no vapor pressure dependence term, so it cannot be used to explain the p/p_{sat} dependence of the capillary force observed for nanoscale systems [82,86–91].

The partial pressure dependence can be explained with the Kelvin equation, which is derived from the thermodynamic equilibrium between the liquid meniscus and the vapor [10]. At constant temperature (T), the free energy of the liquid meniscus is expressed as $\Delta G_i=V_m \cdot \Delta P_L$, where V_m is the molar volume of the liquid. The free energy of the vapor is $\Delta G_v=R_g T \cdot \ln(p/p_{sat})$ (assuming ideal gas behavior), where R_g is the universal gas constant. From $\Delta G_i = \Delta G_v$, the Kelvin radius (a negative number for $0 < p/p_{sat} < 1$) is derived as:

$$r_K = \frac{\gamma V_m}{R_g T} \frac{1}{\ln(p / p_{sat})} \qquad (7.4)$$

Here $\dfrac{\gamma V_m}{R_g T}$, is 0.53 nm for water and 0.51 nm for ethanol at 300 K. At thermodynamic equilibrium, r_e in Equation (7.3) equals r_K.

The meniscus radii, r_m and r_a, can now be related to p/p_{sat} using r_K. In the circular approximation model (Figure 7.13b), $r_m = r_K$ [92]. Then, r_a can be obtained by finding the circle with a radius of r_K that meets the sphere and the flat substrate with known contact angles [92,93]. In reality, r_a and r_m are not constant, and they vary along the meniscus surface (Figure 7.13a). Only r_e is constant at a given temperature and pressure. The exact shape of the liquid meniscus spanning between a sphere and a flat substrate can be obtained by solving the differential Laplace-Young equation, where r_e is set equal to r_K and the liquid contact angles at the sphere and substrate surfaces are used as boundary conditions [93,94]. Once the shape and size of the meniscus is determined, the capillary force can be calculated using Equations (7.1) and (7.2). However, the results of these calculations show very weak partial pressure dependence of the capillary force (F_{cap}), which is inconsistent with experimental data shown in Figure 7.12a [92,93]. The sum of F_{pl} and F_{st} decreases monotonically, but its change is insignificant until p/p_{sat} approaches the saturation point (Figure 7.12b). The calculations confirm that F_{pl} is the main contributing factor to F_{cap}. It appears that the decrease of $-P_L$ upon the increase of p/p_{sat} is almost fully counter-balanced by the increase of the meniscus area in Equation (7.1) [94].

Now, one needs to analyze possible errors in the model. The surface may not be as smooth as a line represented in the simple geometrical models [92]. Modeling the AFM tip as a sphere may not be correct, especially when the tip is dull [87]. More importantly, the solid surface cannot be considered to be a bare, clean solid. The capillary force calculations described previously consider only thermodynamic equilibrium of the meniscus in the "narrow" gap with the gas phase; they do not take into account the equilibrium of the adsorbate layer on the "open" solid surface with the gas phase. As mentioned earlier with the glass example, a solid surface with a high surface energy is readily covered with adsorbed molecules impinging from the gas phase. Without taking into account the presence of the adsorbate layer with which the meniscus and the vapor phase are in equilibrium, the theoretical model cannot represent the real physical system appropriately.

The adsorbate thickness at a given p/p_{sat} can be measured with various techniques, such as vibration spectroscopy, ellipsometry, or quartz crystal microbalance. Figure 7.14a represents the adsorption isotherms of n-propanol and n-pentanol on a clean SiO_2 surface at room temperature measured with attenuated total reflection infrared (ATR-IR) spectroscopy [95]. The thickness of the adsorbed alcohols increases rapidly as the alcohol p/p_{sat} increases from zero to ~10%, where approximately one monolayer is formed. Above this vapor pressure, the adsorbate layer thickness remains relatively constant until the condensation of the liquid layer takes place near the saturation vapor pressure (>90%). This is a typical type-II adsorption isotherm behavior.

The effect of the vapor adsorption isotherm can be accounted for by adding the adsorbate thickness to the solid surface and shifting the meniscus contact point from the solid surface to the adsorbate surface [93,94]. Then, the calculated capillary force for the simple sphere-on-flat geometry shows the strong p/p_{sat} dependence. Figure 7.12c displays the capillary force calculated for a silicon oxide asperity contact covered with alcohol layers formed by adsorption from vapor [93]. The calculation predicts that the capillary force increases as p/p_{sat} increases from zero to ~15% and then decreases as p/p_{sat} increases further. The p/p_{sat} region after the peak F_{cap} is in good agreement with the experimental data [82]. This is the partial pressure

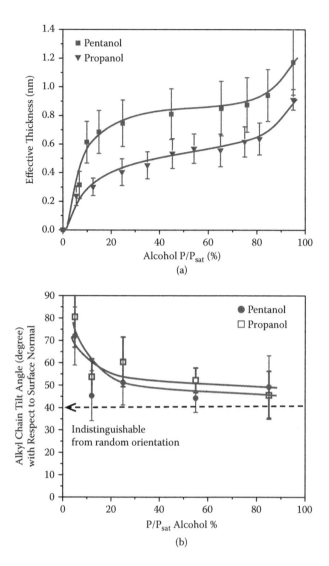

FIGURE 7.14 (a) Adsorption isotherm thicknesses for n-propanol and n-pentanol layers on clean hydrophilic SiO_2 as a function of alcohol partial pressure relative to the saturation pressure. (b) Alkyl chain tilt angles with respect to the surface normal for adsorbed n-propanol and n-pentanol layers as a function of p/psat. (From Barnette, A. L., Asay, D. B., Janik, M. J., and Kim, S. H. 2009. *J. Phys. Chem. C.* 113: 10632.)

regime where the solid surface is fully covered with adsorbed alcohol molecules. In the low pressure region ($p/p_{sat} < 10\%$), the alcohol coverage is incomplete, and there could be a fair amount of direct solid–solid interactions in the measured value. The calculation result shown in Figure 7.12c accounts for the capillary force only and did not include the solid contact contribution. In addition, the alcohol molecules in the incomplete coverage p/p_{sat} regime are tilted toward the surface, while the molecules in the high p/p_{sat} regime are randomly oriented (Figure 7.14b) [95]. Continuum theory calculations cannot predict if the molecular orientation of adsorbate can influence the nano-asperity capillary. In any case, it is interesting to note that, in the experimental data (Figure 7.12a), there seems to be a shoulder at p/p_{sat} ~0.15, which is the pressure where F_{cap} is at its maximum.

Another interesting feature of Figures 7.12a and 7.12c is that a large capillary force is observed for ethanol and that the capillary force decreases as the alkyl chain length of the alcohol increases. Note that the surface tensions of these alcohols are almost the same. In fact, ethanol has a slightly lower surface tension than the others (ethanol = 21.8 mJ/m; propanol = 23.7 mJ/m; butanol = 24.6 mJ/m; pentanol = 24.9 mJ/m). The observed trend is mainly due to the molar volume differences among these alcohol molecules (ethanol = 58.4 cm³/mol; propanol = 74.8 cm³/mol; butanol = 91.5 cm³/mol; pentanol = 108.7 cm³/mol) [82]. The smaller molecule exhibits a larger Laplace pressure at the same partial pressure. This result implies that larger molecules would be more efficient in reducing the adhesion forces when environmental vapors are used to mitigate the adhesion.

It is very important to understand the effect of water adsorption on the asperity adhesion in humid conditions under which water adsorption readily occurs. Figure 7.15 displays the force needed to separate a Si AFM tip (curvature of ~20nm, covered with native oxide) from the clean silicon oxide layer on a Si wafer in humid argon environments [91]. On average, the measured force increases by a factor of ~3 as RH increases from 0 to ~30%, becomes relatively constant within the RH region from 30 to 50%, and then decreases to a value lower than the dry case as RH increases from 50 to 90%.

As in the case of alcohols, the critical information needed to understand this strong RH dependence is the adsorption isotherm behavior of water on the SiO_2 surface. Figure 7.16 displays the OH stretching vibration region of the adsorbed water

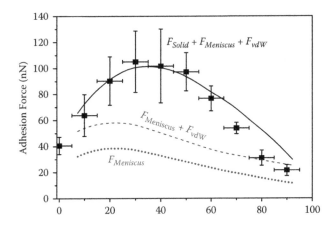

FIGURE 7.15 AFM pull-off force measured for clean hydrophilic SiO2 as a function of relative humidity and calculation results for the contributions from capillary force term ($F_{Meniscus}$), van der Waals term (F_{vdW}), and the ice-ice bridge term (Fsolid). (From Asay, D. B., and Kim, S. H. 2006. *J. Chem. Phys.* 124: 174712.)

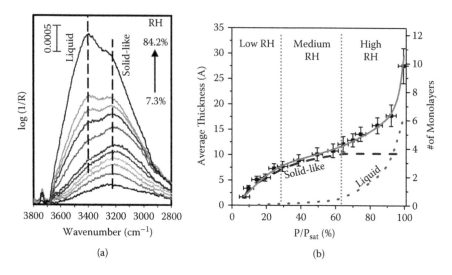

(a)   (b)

FIGURE 7.16 (a) ATR-IR spectra of water adsorbed on silicon oxide at different relative humidities. From lowest to highest log(1/R) signal intensity, relative humidity = 7.3, 9.7, 14.5, 19.4, 24.5, 29.4, 38.8, 49.4, 58.6, 64.3, 69.9, 74.5, 84.2, 92.2, 99.4 %. The O-H stretching vibration peak positions of "ice-like" water and "liquid" water are marked with dotted lines at ~3230 cm–1 and ~3400 cm–1, respectively. (b) Adsorption isotherm of adsorbed water on silicon oxide surface. Square symbols are the total thickness of the adsorbed water layer calculated from the intensity of H-O-H bending vibration peak. The solid line is drawn for visual guidance. The dashed and dotted lines are the thickness of the ice-like water and liquid water layers, respectively. (From Asay, D. B., and Kim, S. H. 2005. *J. Phys. Chem. B* 109: 16760–16763.)

ATR-IR spectra and the adsorption isotherm thickness calculated from the ATR-IR spectra [96]. The average adsorption isotherm thickness appears to follow the typical type-II isotherm curve—fast growth of the thickness in the low humidity region (RH < 20%) and slow growth in the mid-RH region followed by fast growth again in the near-saturation region (RH > 80%). However, notice that the fast growth region extends up to 3~4 layers of water [96]. Another interesting and important fact is that the OH stretch peak position indicates that the adsorbed water layer appears to be solid-like (peak centered at ~3230 cm⁻¹) at low RH and then it becomes covered with a liquid-like structure (peak centered at ~3400 cm⁻¹) as RH increases above 40% [96]. Polarization-dependent IR measurements revealed that the water molecules in the thick layers formed at high RH are randomly oriented, while their orientations deviate from the random orientation in the thin layer formed at the lowest RH [97]. The isosteric heat of adsorption, $q_{st}(\theta)$, was measured to be ~60 kJ/mol at ~6% RH where the solid-like water structure is dominant [98]. This is much larger than the heat of vaporization of liquid water (44 kJ/mol). This value is even larger than the heat of sublimation of ice water (50 kJ/mol). The $q_{st}(\theta)$ value decreases gradually as the adsorbed water layer thickness increases and becomes 43~44 kJ/mol when the total adsorbed layer thickness reaches the region where the liquid structure dominates [98].

With this information about the adsorbed water layer on the silicon oxide surface in humid ambient, the RH dependence of the nano-asperity pull-off force can be explained (Figure 7.15) [91]. When the capillary force is calculated using equations that accurately describe the alcohol capillary behaviour and the liquid property of water, the calculation results underestimate the force needed to pull off the AFM tip in the intermediate RH region [91]. The calculated capillary and van der Waals forces are in good agreement with the experimental data only when RH > 80% where the adsorbed layer behaves like a thick water layer. The discrepancy between the theoretical prediction and the experimental data must be due to the presence of the solid-like water near the solid surface. Although the physical properties of this solid-like structure are not known, one can borrow the properties of ice to estimate the AFM pull-off force. When the force needed to break the contact between solid-like layers at the center of the nano-asperity contact is calculated using the thickness of the solid-like layer determined from the ATR-IR study and added to the total pull-off force, the experimental data can be explained adequately [91].

The examples discussed in this section clearly demonstrate that the environmental effects on adhesion cannot be explained satisfactorily without taking gas adsorption into account. The capillary effect in the alcohol vapor environment shows that both the meniscus surface and the adsorbate layer are in equilibrium with the vapor phase, and the adsorbate thickness should be considered in the force balance to predict the nano-asperity capillary effect properly. The humidity effect on the AFM pull-off force reveals that both the total adsorbate layer thickness and the adsorbate layer structure play important roles.

7.3.2 MITIGATING FRICTION AND PREVENTING WEAR OF SiO₂ IN THE PRESENCE OF ALCOHOL VAPOR

The presence of water and alcohol vapors in the ambient alters the friction and wear behavior of native oxide surfaces on a silicon wafer [99,100]. Figures 7.17a and 7.17b compare the pin-on-disk friction data measured with a 3-mm diameter fused silica ball on a silicon wafer [100]. In an Ar gas with 50% RH, friction coefficients are higher than in the dry case. In 50% p/p_{sat} n-pentanol, the friction coefficient remains low with an average value of 0.1~0.15 for the duration of the experiment for all contact pressures. It is interesting to note that, in an alcohol vapor environment, the friction coefficient measured at the microscale is in the same range [99,101]. In the low load condition (0.1 N), the friction coefficient at 50% p/p_{sat} n-pentanol is slightly higher but much less noisy than that observed in dry Ar. In the dry environment, the presence of cylindrical wear particles produces a third body contact condition, which appears to reduce the contact area and thus reduces friction [102]. In the high load condition (0.7 N), the friction coefficients observed in the 50% p/p_{sat} n-pentanol vapor are lower and fluctuate much less around their average values than those observed in both dry Ar and 50% RH conditions. These friction coefficient data indicate that water vapor is detrimental and n-pentanol vapor is beneficial for attenuating friction and wear.

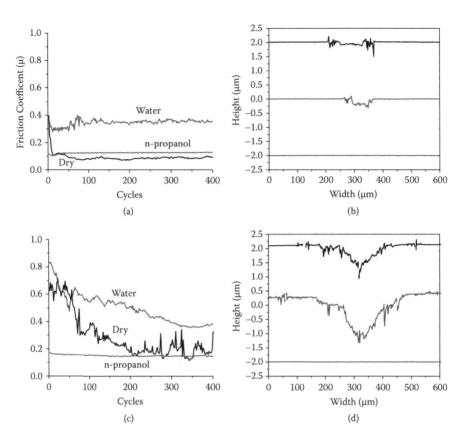

FIGURE 7.17 Friction coefficient measured in dry, 50% RH, and 50% p/psat n-pentanol vapor environments with an applied load of (a) 0.1 N and (c) 0.7 N. Optical profilometry images and characteristic line profiles of wear tracks made in dry, 50% RH, and 50% p/psat n-pentanol vapor environments with an applied load of (b) 0.1 N and (c) 0.7 N. (From Barnette, A. L., Asay, D. B., Kim, D., Guyer, B. D., Lim, H., Janik, M. J., and Kim, S. H. 2009. *Langmuir* 25: 13052.)

Vapor effects are manifested more drastically in the wear behavior, as shown in Figures 7.17b and 7.17d [100]. Compared to the dry Ar case, it can be seen that in 50% RH, the wear volume is increased and the wear pattern is also changed. At 0.1 N load, the wear track made in dry conditions is relatively flat, with some piles of debris on both sides of the wear track. In contrast, the wear track made in humid conditions contains many scratch lines running along the sliding direction, with wear debris particles found mostly at both ends of the wear track. At 0.7 N load, the detrimental effect of humidity is more noticeable. The wear track is much deeper and wider compared to the dry case. In 50% p/p_{sat} n-pentanol, however, no wear of the SiO_2 surfaces was observed as long as the contact load is lower than the fracture strength of the test materials. Since these wear tests were done at the same nominal contact load and the only difference was the vapor condition in the ambient,

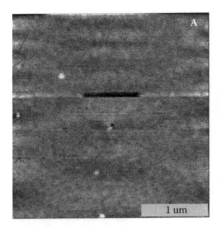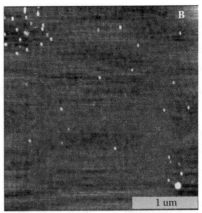

FIGURE 7.18 AFM topography images of the region wear-tested with line scans over 1 μm at 75-nN load in (a) 75% RH and (b) 75% p/p_{sat} n-propanol vapor environments. (From Asay, D. B., Dugger, M. T., Ohlhausen, J. A., and Kim, S. H. 2008. *Langmuir* 24: 155.)

the difference in the wear behavior must be associated with the chemical effect of the adsorbed vapor.

The drastic contrast in the wear behavior, (that is, severe wear in humid environments and no wear in alcohol vapor), is also observed at the nano-scale single asperity sliding. Figure 7.18 displays the AFM images taken at a 3 nN load before and after a 1-μm-long line-scan was for 512 times with a 75 nN applied load (which corresponds to a maximum Hertzian contact pressure of ~3 GPa) [99]. The AFM wear test in 75% p/p_{sat} n-propanol vapor makes no discernable wear trench, while the test in 75% RH humid environment leaves a ~1 nm deep wear trench. The wear prevention capability of the adsorbed n-propanol layer in equilibrium with the gas phase at this high contact pressure condition is truly remarkable.

As shown in Figures 7.14 and 7.16, there are always thin layers of adsorbed molecules on the surface in equilibrium with water in the gas phase. Although strongly bound hydroxyl layers formed on certain mineral surfaces (such as mica) show remarkable lubrication ability for atomically smooth surfaces [103,104], water is not a good lubricant for silicon oxide asperity contacts [105]. It appears that the adsorbed water accelerates the wear of silicon oxide, compared to the dry environment, via some sort of chemical process. By replacing water with alcohol, this chemical wear process seems to be prevented and the silicon oxide surface is protected.

7.3.3 TRIBOCHEMICAL REACTIONS OF ALCOHOL ON SiO$_2$ SURFACES

The fact that friction and wear tests under the same applied load and sliding velocity with the same materials yield very different results depending on the vapor environment implies that the differences must be due to the chemical effects induced by adsorbed molecules, rather than mechanical effects. In order to understand the tribochemistry responsible for lubrication in alcohol vapor environments, one needs

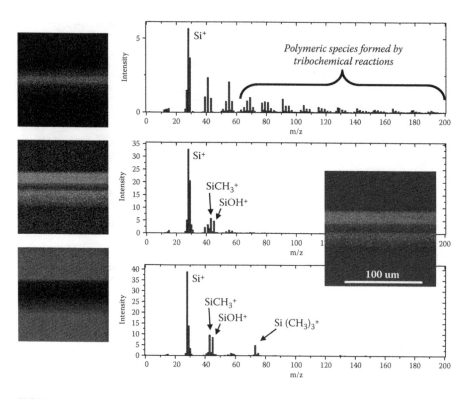

FIGURE 7.19 Principal components resulting from multivariate analysis of ToF-SIMS spectra, showing spatial locations of the components (left) and ion fragmentation spectra (right) obtained for the wear track created on a Si wafer with a 3-mm diameter fused silica ball at a 0.1 N load in 20% p/psat n-pentanol vapor environment. The inset image is the montage of the three component images. (From Kim, S. H., Dugger, M. T., Erdemir, A., Barnette, A. L., Hsiao, E., Marino, M. J., and Eryilmaz, O. L. 2010. *Tribology* 4: 109–114.)

to know the chemical species present in the contact sliding region. Knowing the chemical species produced during the slide from in-situ analysis would be ideal; but this is difficult, if not impossible, to do. The next best option is chemical analysis of the slide track before and after the tribo-test.

The chemical species present before the tribo-test are a monolayer of adsorbed alcohol molecules for partial pressures between 10% and 90% of the saturation vapor pressure, as shown in Figure 7.14. To find the species present in the sliding contact region after the tribo-test, multivariate imaging ToF-SIMS analyses were performed on the wear track on a Si wafer after a pin-on-disc test with a 3-mm diameter fused silica ball. The data shown in Figure 7.19 were obtained after the tribo-test in an Ar environment containing 20% p/p_{sat} n-pentanol vapor at an applied load of 0.1 N [99,106]. The mass spectra of the principal species observed in the wear track are clearly different from those in the vicinity of the wear track. Outside the wear track (bottom panel), typical organic contaminant peaks are observed. Note that the physisorbed n-pentanol cannot be detected in

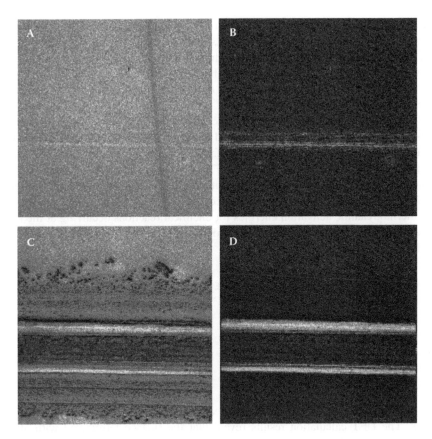

FIGURE 7.20 ToF-SIMS total ion spectra (a, c) and the tribo-product spectra (b, d) of wear regions corresponding to tests at (a, b) 10% p/psat n-pentanol and 0.2 N load and (c, d) 10% p/psat and 0.7 N. The images are 500 μm × 500 μm. (From Barnette, A. L., Dugger, M. T., Ohlhausen, J. A., and Kim, S. H. 2010. *Langmuir* 26: 16299–16304.)

ToF-SIMS since it desorbs when the sample is mounted in vacuum. In the center of the sliding track, where the contact pressure is the highest (top panel), high molecular weight polymeric species are detected instead of chemisorbed pentoxide species. This means that the adsorbed n-pentanol molecules undergo some polymerization reactions under severe tribological contact conditions.

One may think that the tribochemically-produced polymeric species would be responsible for lubrication of the sliding interface. If this were true, then the tribochemical reaction products would still help lubricate the interface upon temporary reduction of the alcohol vapor pressure. However, this is not the case. These polymeric species appear to be associated with, or byproducts of, the wear process. Figure 7.20 shows the correlation between the tribochemical polymer formation and the substrate wear [107]. When the load is low and the alcohol pressure is sufficiently high for effective lubrication, a very small amount of tribochemical polymers is detected. When the lubrication by the adsorbed alcohol is not effective and the substrate gets worn, a larger amount of polymers are formed. The exact reaction mechanisms are not yet fully understood. Alcohols cannot be polymerized via thermal reaction pathways even at high pressure conditions. The tribochemical polymerization may occur due to direct channeling of mechanical energy into the chemical reaction coordinate [108] or the involvement of electrons or ions emitted during the substrate wear [109,110].

If the tribochemical polymerization reaction is not the primary lubrication mechanism, the wear prevention effect observed in the alcohol vapor environment should be the consequence of the monolayer surface chemistry of the adsorbate. We have carried out density functional theory (DFT) calculations for Si–O–Si bond dissociation of the alkoxide and hydroxyl terminated silicon oxide surfaces by reactions with incoming water or alcohol gas molecules [100]. These calculations can give the thermodynamic activation energies for chemical reactions occurring at the surface terminated with various functional groups in the presence of different incoming molecules as shown in Scheme 7.1.

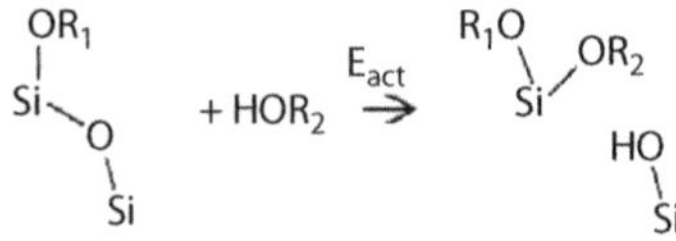

SCHEME 7.1 Reaction between the top layer Si-OR$_1$ (R$_1$ = H or alkyl) with impinging gas molecule HOR$_2$ (R$_2$ = H or alkyl).

The calculated activation energies for reaction Scheme 7.1 are shown in Table 7.2 [100]. The DFT calculation results reveal that the surface terminated with alkyl groups has a much higher activation barrier for the Si–O–Si dissociation reaction than the surface terminated with hydroxyl groups (R$_1$ = H). Furthermore, the activation energy increases as the alkyl group size increases. These findings imply that

TABLE 7.2

Activation Energy for Reactions Shown in Scheme (7.1).

R_1	R_2	Ea (kJ/mol)
H	H	114
CH$_3$	H	151
H	CH$_3$	112
CH$_3$	CH$_3$	154
n-C$_3$H$_7$	n-C$_3$H$_7$	224

Source: Barnette, A. L. Asay, D. B., Kim, D., Guyer, B. D., Lim, H., Janik, M. J., and Kim, S. H. *Langmuir* 25 (2009): 13052.

the SiO_2 surface should be much more wear-resistant in alcohol vapor than in air or water vapor.

7.4   LUBRICATION USING ENVIRONMENTAL EFFECTS

An important example of lubrication by the gas environment is the lubrication of high-current-carrying sliding contacts in electrical motors and generators [111]. In many rotating and linear electrical machines, the brush and slip ring materials and the environment play influential roles affecting their performance. Among various materials, copper and its alloys are used as current-carrying materials. Due to frictional and ohmic heating, electrical contact spots are heated to high temperatures and become susceptible to chemical reactions with the environmental gases. Under typical ambient air conditions, a cuprous oxide (Cu_2O) film is formed at the copper electrical contact surface [111,112]. The condition of this film is often critical for good brush operation. Because of the semiconductor nature of the oxide film, polar effects of the contact voltage drop are often observed [113]. It is important to note that a humid CO_2 atmosphere provides good lubrication to these sliding electrical contacts [114]. Since CO_2 does not chemisorb on copper and its physisorption capability is very weak, it is assumed that the adsorbed water layer on copper facilitates CO_2 adsorption to the electrical contact surface [114]. Compared to the vacuum environment where metal–metal welding can take place at the contact spot, the electrical contact resistance is slightly higher in the humid CO_2 atmosphere due to the presence of oxide layers and adsorbed molecules at the interface. The friction coefficient is much lower and the wear surface is much smoother in the humid CO_2 atmosphere.

Environmental lubrication effects can play an important role in enabling the full degree of motion in microelectromechanical system (MEMS) applications. The first demonstration of MEMS devices with movable structures was made in the 1980's [115–117]. However, commercially successful MEMS devices to date have only either nonmoving parts or contacts whose lateral motion is very restricted [118]. The main challenge in enabling MEMS technology with full degrees of mechanical motions is the friction and wear of the sliding contacts. MEMS are mostly made from silicon using lithographic and micro-machining techniques [118]. However, silicon itself has poor tribological properties and thus is subject to severe wear. Although a number of coating-based lubrication strategies have been proposed and studied extensively, most of them do not last long enough or efficiently prevent wear [118]. Tribological problems associated with MEMS cannot be resolved by applying conventional lubrication methods such as liquid lubricants utilized in the macro-scale. In MEMS, the viscosity of liquid lubricants causes severe power dissipation problems and causes devices to move slowly, canceling one of the principal advantages of micro-scale mechanical machines—that is, low inertia that enables rapid mechanical response.

The lubrication effect of the alcohol vapor discussed in the previous section has been shown to be operational in the MEMS scale. By simply introducing alcohol vapor to the environment, a MEMS device that fails within one or two minutes in dry conditions can be operated continuously for more than two weeks without failure (Figure 7.21) [99,101,106]. Essentially, the MEMS device is being

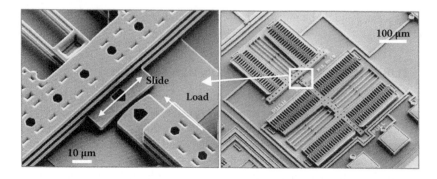

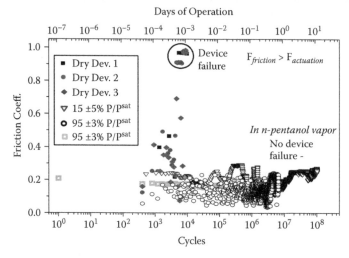

FIGURE 7.21 SEM images of MEMS side-wall tribometer and friction coefficient measured with MEMS tribometer as a function of cycles (bottom axis) and operation time (top axis) in dry nitrogen and n-pentanol vapor environments. (From Kim, S. H., Dugger, M. T., Erdemir, A., Barnette, A. L., Hsiao, E., Marino, M. J., and Eryilmaz, O. L. 2010. *Tribology* 4: 109–114.)

lubricated in the equilibrium vapor adsorption condition. Whenever the bare substrate surface is exposed due to mechanical actions of the sliding interface, the surface can be re-passivated immediately by the impinging gas molecules as long as the partial pressure of the active lubricant molecule in the gas phase is high enough. Similar vapor phase lubrication effects were observed by the J. Krim group using ethanol [119]. The gas adsorption is fully conformal, so even deeply buried surfaces can be lubricated as long as the gap between the surfaces is larger than the mean free path of the lubricant molecule [120]. The thickness of the adsorbed film can be controlled from sub-nm to a few nm by controlling the partial pressure of the lubricant, as the isotherm is a function of the partial pressure (as shown in Figures 7.14 and 7.16). Unlike liquid lubrication, since lubricating molecules are delivered to the interface through the gas phase, there is no power dissipation problem.

7.5 CONCLUDING REMARKS AND FUTURE PROSPECTS

The environmental effects discussed in this review clearly reveal that the friction and wear processes can be accompanied with a variety of chemical reactions. In these chemical reactions, not only the atoms of the contacting or sliding solid surface but also the molecules impinging from the gas or liquid phase are involved. Solid properties alone are not enough to fully understand and control tribological behavior. Tribochemistry is currently at the "qualitative understanding" state; especially when organics are involved, it is still far from molecular-level understanding. A number of factors make the mechanistic study of tribochemistry difficult. First, the tribochemical products involving organics are generated in a mixture form, and their quantity is often too small for separation, purification, and structural analysis. In organic chemistry, nuclear magnetic resonance (NMR) is one of the most-powerful and widely used structural analysis tools. However, it requires at least tens of milligrams of pure samples. It is extremely difficult, if not impractical, to employ a structural characterization tool such as NMR in tribochemical products. Second, the causes or driving forces of the tribochemical reactions are unclear in many cases. Rubbing two solid surfaces can cause a number of physical phenomena. Heat is generated due to dissipation of frictional energy [1]. If the transient temperature at the asperity contact is high enough, thermionic emission of electrons can occur. When the solid surface wears, the bond dissociation process can be accompanied by emission of energetic photons or particles such as electrons, ions, or radicals [110]. This is called tribo-emission. The dangling bonds exposed at the solid surface upon wear will be highly reactive to various organic reactions. Therefore, solid surfaces that are inert in normal reaction conditions can be highly reactive in tribological conditions. In typical chemical reactions, the energy needed to overcome the reaction barrier is provided from a heat source or photons. Since this activation energy can be provided in a controlled way in organic and physical chemistry, one can find and study the molecular energy levels involved in chemical reactions. This "luxury," however, is not available in tribochemical systems; it is very difficult to precisely control the driving force for reactions in tribological systems. In addition, the mechanical force or stress applied to the system could perturb the molecular orbitals [108]; so the use of equilibrium concepts such as simple Lewis acid-base theory may not be adequate to fully understand the tribochemistry, although they could be a reasonable guide [121]. These are some of the key obstacles that one must overcome to bring the field of tribology to molecular-level science and engineering and enable application-driven systems design in tribology. Accomplishing this goal will require the development of more powerful analytical tools and theoretical models to unveil fundamental mechanisms of chemical reactions taking place at tribological interfaces.

ACKNOWLEDGMENTS

The author is grateful for financial support from National Science Foundation (Grant No. CMMI-0408369, 0625493, 1000021) and the Air Force Office of Scientific Research (Grant No. FA9550-08-1-0010).

REFERENCES

1. Heinicke, G. 1984. *Tribochemistry*. Berlin: Akademie-Verlag.
2. Fischer, T. E. 1988. *Ann. Rev. Mater. Sci.* 18: 303–323.
3. Rhee, S. K. 1977. *J. Mater. Sci.* 12: 823–824.
4. Chappell, R. A., and Stoddart, C. T. H. 1977. *J. Mater. Sci.* 12: 2001–2010.
5. Doremus, R. H., Mehrotra, Y., Lanford, W. A., and Burman, C. 1983. *J. Mater. Sci.* 18: 612–622.
6. Wiederhorn, S. M. 1967. *J. Am. Ceram. Soc.* 50: 407–414.
7. Naono, H., Fujiwara, R., Yagi, and M. J. Coll. 1980. *Interface Sci.* 76: 74–82.
8. Somorjai, G. A. 1994. *Introduction to Surface Chemistry and Catalysis*. New York: John Wiley & Sons.
9. Yang, R. R. 1997. *Gas Separation by Adsorption Processes*. River Edge, NJ: Imperial College Press.
10. Adamson, A. W. 1990. *Physical Chemistry of Surfaces*, 5th ed. New York: John Wiley & Sons.
11. Zhou, X.-L.., Zhu, X.-Y., and White, J. M. 1991. *Surf. Sci. Rep.* 13: 73–220.
12. Beyer, M. K., and Clausen-Schaumann, H. 2005. *Chem. Rev.* 10: 2921–2948.
13. Boldyrev, V. V. 2006. *Russ. Chem. Rev.* 75: 177–189.
14. Buckley, D. H. 1978. *Wear* 46: 19–53.
15. McFadden, C., Soto, C., and Spencer, N. D. 1997. *Tribol. Int'l.* 30: 881–888.
16. Martin, J.-M., Le Mogne, T., Boehm, M., and Grossiord, C. 1999. *Tribol. Int'l* 32: 617–626.
17. Gellman, A. J., and Ko, J. S. 2001. *Tribol. Lett.* 10: 39–44.
18. Lara-Romero, J., Maya-Yescas, R., Luis Rico-Cerda, J., Luis Rivera-Rojas, J., Castillo, F. C., Kaltchev, M., and Tysoe, W. T. 2006. *Thin Solid Films* 496: 463–468.
19. Special issue of MRS Bulletin on "In Situ Tribology, 2008, vol. 33, 1145–1189.
20. Buckley, D. H. 1968. *Wear* 11: 405–419.
21. Buckley, D. H. 1968. *J. Appl. Phys.* 39: 4224–4233.
22. Buckley, D. H. 1972. *Wear* 20: 89–103.
23. Wheeler, D. R. 1976. *J. Appl. Phys.* 47: 1123–1130.
24. Mishina, H. 1992. *Wear* 152: 99–110.
25. Goto, H., and Buckley, D. H. 1991. *Wear* 143: 15–28.
26. Boehm, M., Martin, J. M., Grossiord, C., and Le Mogne, T. 2001. *Tribol. Lett.* 11: 83–90.
27. McFadden, C. F., and Gellman, A. J. 1997. *Surf. Sci.* 391: 287–299.
28. McFadden, C. F., and Gellman, A. J. 1998. *Surf. Sci.* 409: 171–182.
29. Niebuhr, D. 2007. *Wear* 263: 88–92.
30. Wu, X., Cong, P., Nanao, H., Minami, I., and Mori, S. 2004. *Tribol. Lett.* 17: 925–930.
31. Singer, I. L., Le Mognoe, T., Donnet, C., and Martin, J. M. 1996. *J. Vac. Sci. Technol.* A14: 38–45.
32. Lara, J., and Tysoe, W. T. 1998. *Langmuir* 14: 307–312.
33. Gao, F., Furlong, O., Kotvis, P. V., and Tysoe, W. T. 2005. *Tribol. Lett.* 20: 171–176.
34. Graham, E. E., and Klaus, E. E. 1986. *ASLE Trans.* 29: 229.
35. Graham, E. E., Nesarikar, A., Forster, N., and Givan, G. 1993. *Lubr. Eng.* 49: 713.
36. Sinfelt, J. H. 1973. *AIChE J.* 19: 673–683.
37. Bligaard, T., and Nørskov J. K. Heterogenous catalysis. 2007. In *Chemical Bonding at Surfaces and Interfaces*, ed. A. Nilsson, L. Pettersson, J. Norskov. New York: Elsevier.
38. Hiratsuka, K., Abe, T., and Kajdas, C. 2010. *Tribol. Int'l.* 43: 1659–1664.
39. Shinchi, A., Imada, Y., Honda, F., and Nakajima, K. 1999. *Wear* 230: 78–85.
40. Paul, E., Evans, C. J., Mangamelli, A., McGlauflin, M. L., and Polvanib, R. S. 1996. *Precision Eng.* 18: 4–19.

41. Mori, S., and Shitara, Y. 1994. *Appl. Surf. Sci.* 78: 269–273.

42. Wu, X., Kobayashi, N., Nanao, H., and Mori, S. 2005. *Tribol. Lett.* 18: 239–244.

43. Fischer, T. E., and Tomizawa, H. 1985. *Wear* 105: 29.

44. Tomizawa, H., and Fischer, T. E. 1986. *ASLE Trans.* 29: 481–488.

45. Tomizawa, H., and Fischer, T. E. 1987. *ASLE Trans.* 30: 41–46.

46. Chen, M., Kato, K., and Adachi, K. 2001. *Wear* 250: 246–255.

47. Iler, R. K. 1979. *The Chemistry of Silica,* p. 622. New York: Wiley.

48. Muratov, V. A., Olsen, J. E., Gallois, B. M., Fischer, T. E., and Bean, J. C. 1998. *J. Electrochem. Soc.* 145: 2465–2470.

49. Sugita, T., Ueda, K., and Kanemura, Y. 1984. *Wear* 97: 1.

50. Basu, B., Vitchev, R. G., Vleugels, J., Celis, J. P., and van der Biest, O. 2000. *Acta Mater.* 48: 2461–2471.

51. Wallbridge, N., Dowson, D., and Roberts, E. W. 1983. In *Wear of Materials,* ed. K. C. Ludema. New York: ASME.

52. Chen, C. P., and Knapp, W., J. J. 1977. *Amer. Chem. Soc.* 60: 87–104.

53. Hibi, Y., and Enomoto, Y. 1989. *Wear* 133: 133–145.

54. Kim, B. I., Lee, S., Guenard, R., Torres, L. C. F., Perry, S. S., Frantz, P., and Didziulis, S. V. 2001. *Surf. Sci.* 481: 185–197.

55. Cong, P., Kobayashi, K., Li, T., and Mori, S. 2002. *Wear* 252: 467–474.

56. Wu, X., Cong, P., Nanao, H., Kobayashi, K., and Mori, S. 2002. *Langmuir* 18: 10122–10127.

57. Lauer, J. L., Blanchet, T. A., Vleck, B. L., and Sargent, B. 1993. *Surf. Coat. Technol.* 62: 399–405.

58. Nakayama, K., and Hashimoto, H. 1996. *Tribol. Int'l.* 29: 385–393.

59. Savage, R. H., and Schaeffer, D. L. 1956. *J. Appl. Phys.* 27: 136.

60. Yen, B. K. 1996. *Wear* 192: 208–215.

61. Yen, B. K., Schwickert, B. E., and Troney, M. F. 2004. *Appl. Phys. Lett.* 84: 4702–4704.

62. Chandrasekar, S., and Bhushan, B. 1992. *Wear* 153: 79–89.

63. Miyoshi, K., Wu, R. L. C., Garscadden, A., Barnes, P. N., and Jackson, H. E. 1993. *J. Appl. Phys.* 74: 4466–4454.

64. Perry, S. S., Ager, J. W., and Somorjai, G. A. 1993. *J. Mater. Res.* 8: 2577–2586.

65. Gardos, M. N. 1999. *Surf. Coat. Technol.* 113: 183–200.

66. Robertson, J. 1999. *MRS Symp. Proc.* 555: 12.

67. Grill, A. 1993. *Wear* 143–153.

68. Grill, A. 1997. *Surf. Coat. Technol.* 94–95: 507–513.

69. Kim, H. I., Lince, J. R., Erylmaz, O. L., and Erdemir, A. 2006. *Tribol. Lett.* 21: 53–58.

70. Heimberg, J. A., Wahl, K. J., Singer, I. L., and Erdemir, A. 2001. *Appl. Phys. Lett.* 78: 2449–2451.

71. Borodich, F. M., and Keer, L. M. 2005. *Thin Solid Film* 476: 108–117.

72. Kim, D. S., Fischer, T. E., and Gallois, B. 1991. *Surf. Coat. Technol.* 49: 537–542.

73. Wu, X., Ohana, T., Tanaka, A., Kubo, T., Nanao, H., Minami, I., and Mori, S. 2007. *Diamond & Rel. Mater.* 16: 1760–1764.

74. Fontaine, J., Donnet, C., Grill, A., and LeMogne, T. 2001. *Surf. Coat. Tech.* 146–147: 286–291.

75. Konca, E., Cheng, Y.-T., Weinerb, A. M., Daschc, J. M., and Alpasa, A. T. 2007. *Tribol. Trans.* 50: 178–186.

76. Bhushan, B., and Ruan, J. 1994. *Surf. Coat. Technol.* 68/69: 644–650.

77. Jesh, M. S., and Segar, P. R. 1999. *Tribol. Trans.* 42: 310–316.

78. Strom, D., Bogy, D., Walmsley, R. G., Brandt, J., and Bhatia, C. S. 1994. *J. Appl. Phys.* 76: 4651–4655.

79. Donnet, C., Martin, J. M., Mogne, T. L., and Belin, M. 1996. *Tribol. Int'l.* 29: 123–128.

80. Voevodin, A. A., Fitz, T. A., Hum, J. J., and Zabinski, J. S. 2002. *J. Vac. Sci. Technol. A* 20: 1434–1444.

81. Zabinski, J. S., Bultman, J. E., Sanders, J. H., and Hu, J. J. 2006. *Tribol. Lett.* 23: 155–163.

82. Asay, D. B., and Kim, S. H. 2007. *Langmuir* 23: 12174.

83. Ramirez-Aguilar, K. A., and Rowlen, K. L. 1998. *Langmuir* 14: 2562–2566.

84. Israelachvili, J. N. 1992. *Intermolecular and Surface Forces*, 2nd ed. San Diego, CA: Academic Press.

85. McFarlane, J. S., and Tabor, D. 1950. *Proc. Royal Soc. London* A202: 224.

86. Binggeli, M., and Mate, C. M. 1994. *Appl. Phys. Lett.* 65: 415.

87. Xiao, X., and Qian, L. 2000. *Langmuir* 16: 8153.

88. Jones, R., Pollock, H. M., Cleaver, J. A. S., and Hodges, C. S. 2002. *Langmuir* 18: 8045.

89. He, M., Blum, A. S., Aston, D. E., Buenviaje, C., and Overney, R. M. 2001. *J. Chem. Phys.* 114: 1355.

90. Nosonovsky, M., and Bhushan, B. 2008. *Phys. Chem. Chem. Phys.* 10: 2137.

91. Asay, D. B., and Kim, S. H. 2006. *J. Chem. Phys.* 124: 174712.

92. Butt, H.-H., and Kapple, M. 2009. *Adv. Colloid Interface Sci.* 146: 48.

93. Asay, D. B., de Boer, M. P., and Kim, S. H. 2010. *J. Adh. Sci. Technol.*, 24: 2363–2382.

94. Hsiao, E., Marino, M. J., and Kim, S. H. *J. Coll. 2010. Interface Sci.* 352: 549–557.

95. Barnette, A. L., Asay, D. B., Janik, M. J., and Kim, S. H. 2009. *J. Phys. Chem. C.* 113: 10632.

96. Asay, D. B., and Kim, S. H. 2005. *J. Phys. Chem. B* 109: 16760–16763.

97. Barnette, A. L., Asay, D. B., and Kim, S. H. 2008. *Phys. Chem. Chem. Phys.* 10: 4981.

98. Asay, D. B., Barnette, A. L., and Kim, S. H. 2009. *J. Phys. Chem. C.* 113: 2128.

99. Asay, D. B., Dugger, M. T., Ohlhausen, J. A., and Kim, S. H. 2008. *Langmuir* 24: 155.

100. Barnette, A. L., Asay, D. B., Kim, D., Guyer, B. D., Lim, H., Janik, M. J., and Kim, S. H. 2009. *Langmuir* 25: 13052.

101. Asay, D. B., Dugger, M. T., and Kim, S. H. 2008. *Tribol. Lett.* 29: 67.

102. Fischer, T. E., Zhu, Z., Kim, H., and Shin, D. S. 2000. *Wear* 245: 53–60.

103. Raviv, U., and Klein, J. 2002. *Science* 297: 1540–1543.

104. Raviv, U., Laurat, P., and Klein, J. 2001. *Nature* 413: 51–54.

105. Helt, J. M., and Batteas, J. D. 2005. *Langmuir* 21: 633–639.

106. Kim, S. H., Dugger, M. T., Erdemir, A., Barnette, A. L., Hsiao, E., Marino, M. J., and Eryilmaz, O. L. 2010. *Tribology* 4: 109–114.

107. Barnette, A. L., Dugger, M. T., Ohlhausen, J. A., and Kim, S. H. 2010. *Langmuir* 26: 16299–16304).

108. Konôpka, M., Turanský, R., Reichert, J., Fuchs, H. Marx, D., and Štich, I. 2008. *Phys. Rev. Lett.* 100: 115503.

109. Nakayama, K., and Hashimoto, H. 1996. *Tribol. Int'l* 29: 385–393.

110. Nakayama, K., and Martin, J. M. 2006. *Wear* 261: 235–240.

111. McNab, I. R. 2980. *Wear* 59: 259–276.

112. Furlong, O., Li, Z., Gao, F., and Tysoe, W. T. 2008. *Tribol. Lett.* 31: 167–176.

113. Belyi, V. A., Konchits, V. V., and Savkin, V. G. 1982. *Wear* 78: 249–258.

114. Singh, B., Zhang, J. G., Hwang, B. H., and Vook, R. W. 1982. *Wear* 78: 17–28.

115. Fan, L.-S., Tai, Y. C., and Muller, R. S. 1989. *Sens. Actuators* 20: 41.

116. Trimmer, W. S. N., and Gabriel, K. J. 1987. *Sens. Actuators* 11: 189.

117. Fan, L.-S., Tai, Y.-C., and Muller, R. S. 1988. *IEEE Trans. Electron Devices* 35: 724.

118. Kim, S. H., Asay, D. B., and Dugger, M. T. 2007. *NanoToday* 2: 22–29.

119. Hook, D. A., Miller, B. P., Vlastakis, B. M., Dugger, M. T., and Krim, J. (submitted).

120. Gellman, A. J., 2004. *Tribol. Lett.* 17: 455.

121. Fischer, T. E., and Mullins, W. M. 1992. *J. Phys. Chem.* 96: 5690–5701.

8 Molecular Dynamics Simulation of Nanotribology

Ashlie Martini

CONTENTS

8.1 INTRODUCTION

The objectives of this chapter are to provide the reader with awareness of the opportunities and limitations of molecular dynamics (MD) simulation for nanotribology research, background that enables interpretation of MD studies reported in the literature, and an understanding of the important contributions made to nanotribology through MD-based research. The chapter will be divided into three sections: an introduction to what molecular dynamics simulation is and what tribological phenomena can be studied using it, and two sections that focus on areas MD research has already contributed to significantly, namely atomic-scale friction and thin film lubrication. This is not meant to be a comprehensive review of the literature, and the reference list is by no means complete. The intention is to provide an overview of the contributions of MD to this field, and references are cited either as examples of a simulation method or result, or to direct the reader to a previous study that may provide more detail on a given topic.

The first question is as follows: when do we need atomic-scale simulation? The simple answer is that we need atomistic simulation to study phenomena that cannot be fully understood from experimental measurements and cannot be captured accurately using continuum assumption-based models (Robbins and Smith 1996). However, it is much more difficult to identify exactly the length scale at which this transition occurs. It has been proposed that, for tribology studies, the transition between atomistic and continuum occurs when we cannot define the size of the contact area between two surfaces; contact area is a relatively straightforward concept on larger length scales, but becomes somewhat ambiguous on the atomic scale. One proposed solution is to define the contact area as the region that circumscribes all atomic bonds between two materials (Mo et al. 2009). But even this is somewhat subjective given that the existence of a bond is defined by the model chosen to describe the interactions between the atoms. Another proposal is that atomistic modeling should be employed when continuum assumption-based material property models can no longer make accurate predictions. For example, density is homogeneous in a bulk fluid, but may exhibit anisotropic behavior when a fluid is confined to molecular length scales (Hu and Granick 1998). Similarly, continuum models cannot typically capture atomic-scale phenomena such as the dependence of friction on atomic-scale surface inhomogenities, nonlinear elastic deformations of the interface, and contributions of plasticity to sliding-induced energy dissipation (Szlufarska et al. 2008).

Once it is determined that atomistic modeling is necessary, a computational tool frequently employed is MD simulation. MD simulation is a powerful modeling tool that is capable of capturing the complex, nonzero temperature dynamics typical of tribological phenomena. MD is a computational tool used to describe how positions, velocities, and orientations of molecules change over time. The simulated time-evolution of the system is based on empirical potential energy models that describe the interaction between atoms. These models relate energy (or force) to the positional configuration of the system, which can in turn be used to calculate particle acceleration via Newton's Second Law. Numerical integration is performed to calculate particle velocities, and then each particle is moved through a distance equal to its calculated velocity multiplied by the simulation time step. Thus, MD is used as a computational *experiment* where a system is defined, allowed to evolve, and then observations are made based on its evolution.

Molecular dynamics simulation has been employed to study all aspects of tribology (contact, adhesion, friction, wear, and lubrication) to varying degrees. The use of MD to study wear has been somewhat limited due to the inability of many empirical models to accurately describe the formation and breaking of covalent bonds, especially for large systems. Although models that can capture chemical reactions have been effectively employed (see, for example, Gao et al. 2002), these methods are typically restricted to a few atomic species and relatively computationally expensive, and so have not been used by a wide range of researchers. One aspect of wear that has been studied extensively using MD is nanoindentation of metals (see, for example, Landman et al. 1992 for Si/Si contacts and Nieminen et al. 1992 for Cu/Cu contacts). The typical procedure is to bring one surface into contact with another, stop at some point, and then pull the two surfaces apart. The resultant force–displacement curve lends insight into how the materials wear (e.g., plastic deformation or atoms transferred from one surface to the next) due to the indentation process. Figure 8.1 illustrates characteristic results from this type of simulation for contact between platinum and gold. A thorough review of MD studies of nanoindentation is available elsewhere (Szlufarska et al. 2008). Although the application of MD to study wear has been somewhat limited compared to MD-based studies of friction and lubrication,

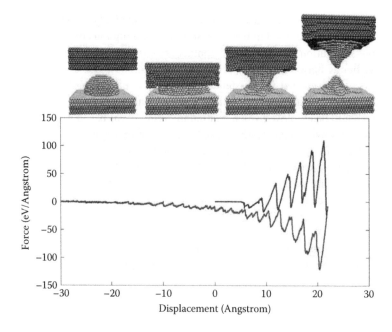

FIGURE 8.1 Force-displacement curve resulting from an MD simulation of nanoindentation where one surface is brought into contact with the other and then pulled away. Above the plot are snapshots of the model before, during, and after indentation.

it is expected that the current development of more and better empirical models will enable rapid growth of this research area.

The friction and lubrication aspects of nanotribology have already been studied extensively using MD simulation. Specifically, MD has been used to make significant contributions to understanding the fundamental mechanisms of atomic-scale friction and thin film lubrication. These two phenomena will be discussed in detail in the following sections.

8.2 ATOMIC-SCALE FRICTION

Atomic-scale friction can be loosely defined as resistance to motion due primarily to atomic interactions (as opposed to larger-scale surface features). This phenomenon has been extensively studied using MD because its atomic scale and dynamic nature are well suited to this type of modeling and because model predictions can be compared directly to experimental data obtained using methods such as atomic force microscopy (AFM). Because the number of atoms that can be modeled is limited, an MD simulation of atomic-scale friction typically consists of only the apex of an AFM probe tip in contact and a substrate that is made effectively infinite by imposing periodic boundary conditions. A representative snapshot of one such model is shown in Figure 8.2, where the spheres represent metal atoms, in this case platinum on gold.

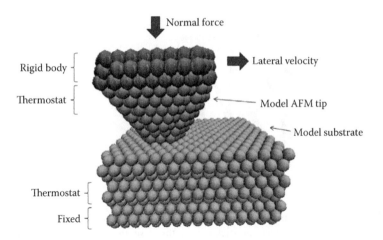

FIGURE 8.2 Atomistic model of the apex of an atomic force microscope tip moving laterally across a substrate.

Although there are many variations on the details of this model, some features are sufficiently standard at this point that they can be presented here. As shown in Figure 8.2, the atoms in the tip–substrate model are not all treated the same during the simulation. The lowermost layers of atoms are fixed in order to model rigidity that would be introduced by the bulk of the substrate that is not explicitly modeled. It is important, of course, to include enough nonrigid atoms to avoid artifacts induced by the rigid layers. Atoms in the middle regions of both the model tip and substrate are mobile, but the Newtonian dynamics are altered by application of a thermostat that effectively removes heat from the system. As mentioned before, the number of atoms that can be modeled is limited, and so heat generated in the contact area that would be conducted away from the interface in the physical system must be removed numerically via the thermostat. It is notable that a thermostat is not applied near the contact area to minimize the effect it might have on frictional behavior. Also, for fast sliding speeds, the thermostat may need to be modified to exclude the contribution of the lateral tip motion to atomic velocity. The atoms near the interface in both the tip and substrate are unrestricted such that their positions evolve according to Newtonian dynamics. The uppermost atomic layers of the model tip are typically fixed relative to each other and so move as a rigid body. It is to this group of atoms that a normal load or lateral velocity is applied to the system. Measurements obtained from an AFM experiment will necessarily be a function of the compliance inherent in the instrument's components during scanning. Compliance can be introduced into the model by incorporating virtual atoms that do not experience interatomic interactions, but are coupled to real atoms in the top layers of the tip through harmonic springs. During the simulation, the virtual atoms are uniformly translated laterally at a constant scanning velocity, and the tip moves through the interactions between the real atoms and virtual atoms, thus reproducing the compliance that exists in the physical experiment.

The characteristic result of both an experiment and an MD simulation of sliding contact is a friction trace—a plot of the lateral force between the two sliding

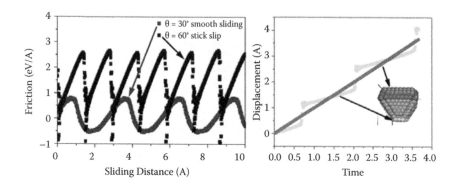

FIGURE 8.3 (a) Friction traces predicted by MD simulation illustrating the difference between smooth sliding and atomic stick-slip. In this case, the transition between these two friction regimes is caused by rotation of the tip relative to the substrate. (b) Displacement of the top part of a tip that is moving at constant velocity and the bottom part of the tip that is in contact with the substrate illustrating stick-slip motion.

materials as a function of either time or sliding distance. A representative force trace from an MD simulation of a copper–copper system is shown in Figure 8.3a. Atomic-scale friction is characterized in terms of both the magnitude and "shape" of the friction trace. Specifically, a friction trace is characterized by the mean friction, maximum friction (also called the *static friction*), and transitions between smooth sliding (tip moves continuously), single-stick-slip (tip repeatedly sticks and then slips a distance of one lattice spacing), and multiple-stick-slip (slip distance is greater than one lattice spacing). The mean friction, the time average of the friction trace, and the maximum friction, the average value of the peak friction, are particularly important parameters because they can be compared to experimental measurements more easily than other frictional properties. Another characteristic of research interest is the transition between smooth sliding and the eponymous atomic stick-slip. As the name implies, in atomic stick-slip, the movement of the two surfaces relative to each other is not steady; instead they *stick* together and then *slip* relative to each other. This behavior is reflected in both the movement of the atoms in the tip as shown in Figure 8.3b and in the "sawtooth" pattern of the friction trace as shown in Figure 8.3a. In contrast, Figure 8.3a also shows a representative friction trace from smooth sliding in which the friction varies but does not exhibit the sharp drops caused by discrete slip events.

Atomic-scale friction is significantly affected by many parameters, including load, velocity, contact area, commensurability, and temperature. These dependencies have been studied extensively using MD, and each will be discussed in the following sections. Note that most of these trends have also been observed experimentally; however, only the MD contributions will be discussed here.

At the macroscopic scale, load directly affects friction force, and the two are linearly related via the friction coefficient. Although the trend is the same on the atomic scale, the friction–load relationship is much more complex because load affects atomic friction both through changes in the contact area and the surface

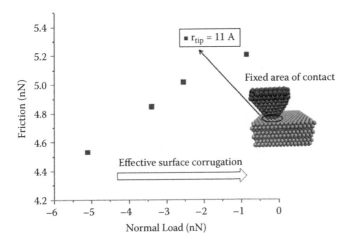

FIGURE 8.4 Increase of friction with normal load where the influence of the contact area has been intentionally excluded to highlight the effect of surface energy corrugation.

energy "corrugation." *Surface energy corrugation* is the distribution of potential energy of a surface, which an atom moving over that surface would "feel" via interatomic interactions. This characteristic of a surface is relevant to atomic stick-slip because it defines the effective height of the potential energy barrier that interface atoms must overcome before they can slip in the direction of the imposed velocity. This effect can be isolated from the effect of contact area using MD by designing a model tip with a constant contact area as shown in Figure 8.4. It can be observed that friction increases (albeit somewhat moderately) with increasing load due to the effect of surface corrugation.

To understand the role of contact area in the load dependence of friction requires that we first define the concept of an atomic-scale contact area. On the macroscopic scale, the term *contact area* is unambiguous, and Hertz theory tells us the contact radius should increase with the effective radius of the contacting bodies, load, and effective elastic compliance of the contact. However, the relationship between these parameters is much more complicated on the atomic scale. The most commonly used method for estimating contact area is based on calculation of the number of surface atoms with nonzero normal force. This number is then converted to a contact radius by dividing the total number of surface atoms and multiplying by the area enclosing the contacting surface atoms (Luan and Robbins 2005). This method has been employed to illustrate the failure of the Hertz theory at atomic length scales by modeling hemispherical tips with the same effective radius but different atomic surface structures. Elastic indentation studies of these three model tips revealed that they exhibited very different contact areas and stress distributions (Luan and Robbins 2005). To isolate the effect of contact area on friction, one can model a tip with a flat surface with a constant number of atoms at the interface. As shown in Figure 8.5 for platinum on gold, increasing contact area results in a slightly sublinear increase of both the mean and maximum (not shown) friction. However, to this point, a general theory for this relationship has not been developed. The overall effect of load on

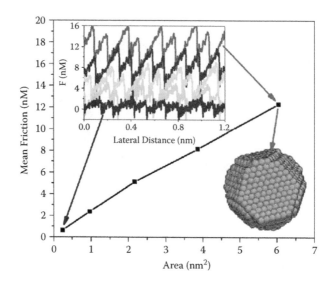

FIGURE 8.5 Mean friction force as a function of contact area illustrating the slightly sublinear relationship. Inset: Friction traces corresponding to the five self-similar model tips with different numbers of interface atoms.

friction (including both surface corrugation and contact area effects) can be evaluated using a hemispherical model tip. As one example, MD simulation of a hemispherical diamond tip with a radius of approximately 1.5 nm showed an increase in friction of ~20 nN in response to a ~125 nN increase in load (Gao et al. 2007).

Another interface property that can significantly affect atomic-scale friction is commensurability. Two surfaces are said to be commensurate if they share common periodicities. Consider the two simplified surfaces shown in Figure 8.6 having lattice constants a and b. The surfaces will be commensurate if the ratio of a to b is equal to the ratio of M to N where M and N are integers; otherwise, they are incommensurate. In practice, this definition is not completely rigid since two crystalline materials with a similar (but not the same) lattice constant may behave as commensurate due to coherency strain. Although it is now well accepted that commensurability has a significant effect on atomic friction, the details of this relationship are still under investigation. In one example, a simulation of Lennard-Jones spheres predicted that friction will be proportional to load for commensurate systems, but that friction and load will be related via a power law with exponent ~0.63 for incommensurate surfaces (Wenning and Müser 2001). Figure 8.7 illustrates the effect of commensurability for contact between metals with similar lattice constants (platinum and gold). As the surfaces are rotated relative to one another, the contact varies between effectively commensurate and incommensurate, causing the mean friction to change significantly (~8 nN in this case). It is also believed that commensurability plays a major role in determining whether a contact exhibits smooth sliding or atomic stick-slip.

Perhaps the most frequently studied effect in atomic-scale friction research is that of velocity. Experiments, analytical models, and MD simulations have shown that

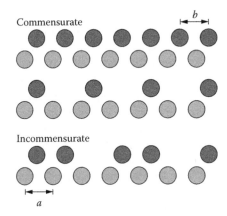

FIGURE 8.6 Heuristic illustration of commensurability between two crystalline surfaces.

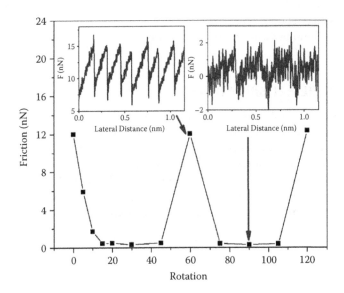

FIGURE 8.7 Results from a simulation of atomic friction in which one crystalline material is rotated relative to the other to vary the contact between commensurate and incommensurate.

friction increases with velocity, and this relationship is almost always reported to be logarithmic. It has been proposed that this dependence is due to phonon excitation that promotes continuing slip events, and to the delay between the movement of atoms to which lateral force or velocity is applied and atoms in the contact (Sørensen et al. 1996). While it is commonly accepted that friction increases with velocity, the rate of this increase is, at this time, still a matter of debate. Most simplified, Tomlinson-type models predict the velocity dependence parameter, λ, to be 2/3 (i.e., friction increases with the natural log of velocity to the two-thirds power). However, most MD simulations are run over such a small range of velocities that the data could be made to fit the expected trend with multiple values of λ. The primary issue in predicting velocity dependence using MD at this point is that simulations are typically restricted to very large velocities. Standard MD necessarily uses a very small time step (~fs), which limits the total simulated duration (~ns) and therefore requires unrealistically large sliding speeds (~m/s) in order to observe most physical phenomena. However, some researchers have recently employed methods in which simulations are run parallel in time, allowing access to slower sliding speeds. Such methods (for example, parallel replica dynamics [Voter 1998]) are based on the assumption that the system being modeled advances to new configurations via infrequent events and that there are well-defined instantaneous rate constants for these events that are independent of the driving rate. This approach has recently been applied to study atomic stick-slip friction of Lennard-Jones spheres (Kim and Falk 2010) and pure metals (Mishin et al. 2007; Martini et al. 2009; Perez et al. 2010). An example of results obtained for a copper–copper system using this method at velocities as low as 0.001 m/s is shown in Figure 8.8. At the time of this writing, the slowest speed for which atomic friction been reported is 2.5 mm/s. The disconnect between simulated and experimental sliding speeds has recently been shown, along with mass difference, to be a limiting factor in direct comparison between predicted and measured results (Li et al., 2011).

8.3   THIN FILM LUBRICATION

The term thin film lubrication (TFL) has multiple meanings, depending on the research community. Some say it is the regime between boundary lubrication and elastohydrodynamic lubrication that exhibits behavior different from both. In other research communities, any lubricant with thickness on the order of nanometers is considered to be a thin film lubricant. Alternatively, it might be said that TFL describes lubrication where atomic-scale phenomena become significant. This last definition is the one we will use in this section. MD simulation is

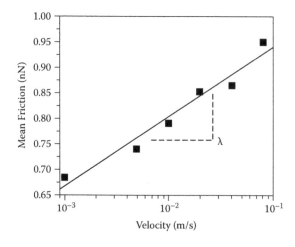

FIGURE 8.8 Mean friction increasing logarithmically with velocity (where λ is the velocity dependence parameter) obtained using Parallel Replica Dynamics enabling simulated sliding speeds lower than standard MD.

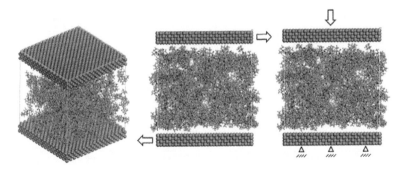

FIGURE 8.9 Atomistic model of thin film lubrication. (a) Perspective view. (b) 2D profile illustrating the fluid under shear, and (c) 2D profile illustrating the fluid in compression

an appropriate tool to model these atomic-scale phenomena. The unique behavior of TFL can, for the most part, be tied to fluid material properties and behaviors that are different when a fluid is confined to molecular length scales than when it is in bulk. MD has been used to study TFL in terms of characteristics including density, solvation pressure, viscosity, and interface slip. An MD model of TFL typically consists of fluid molecules confined in a molecular scale channel by two solid walls, an example of which is shown in Figure 8.9 (in this case polyphenylether).

One of the first properties of confined fluids to be studied using atomistic simulation was anisotropic density. It was shown more than thirty years ago using Monte Carlo simulation of Lennard-Jones spheres that confined fluid atoms tend to orient themselves in layers parallel to the confining walls (Snook and van Megen 1979). This has since been observed many times over using MD modeling (see, for example, Thompson and Robbins 1990). The molecular layering effect is typically characterized as a density profile that can be calculated from the time-averaged position of the atoms relative to the walls. The fluid region of the simulation cell is divided into many thin layers of constant volume, referred to as *bins*. Then, for each simulation time step, the number of atoms in each bin is counted. These values, averaged over time, yield a density profile such as that shown in Figure 8.10a for n-decane. The anisotropy of thin films can also be characterized in terms of an orientation parameter, S, that gives insight into how molecules are aligned relative to the confining surfaces. A typical definition is $S(y)=(3<\cos^2\theta>-1)/2$, where y is the center of mass of the chain molecule between the confining walls, and θ is the angle between the molecular axis vector and the y-axis such that $-0.5 \leq S \leq 1.0$. A value of -0.5 indicates that the molecules are parallel to the walls, 1.0 indicates molecules are perpendicular to the walls, and 0 corresponds to no consistent ordering. It has been observed that molecules in near wall fluid layers tend to orient parallel to the walls ($S \approx -0.5$) while the fluid in the central region is isotropic ($S \approx 0$) (Gupta et al. 1997). An example of this behavior is illustrated in Figure 8.10b.

A property closely related to anisotropic density is compressibility. Compressibility can be studied using MD by fixing one of the confining walls and applying a constant normal load to the other as illustrated in Figure 8.9c. The resultant change in

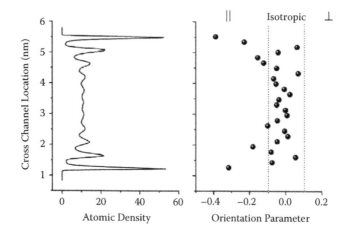

FIGURE 8.10 The (a) local density and (b) orientation parameter as functions of position relative to the confining walls illustrating anisotropy of molecularly confined films.

density (or equivalently film thickness for a simulation of constant wall area) is a measure of the material compressibility. Figure 8.11 illustrates the response of four fluids with different molecular structures to compression. It can be observed that the steady-state film thickness is smaller (density is greater) at larger pressures, and that the molecular structure of the fluid molecules affects this relationship. Thin film compressibility can also be evaluated in terms of its bulk modulus, a measure of resistance to compression. The bulk modulus can be calculated from MD by increasing the pressure a small amount and measuring the resultant change in volume. The bulk modulus is then the ratio of the pressure change to the volume change multiplied by the negative of the initial volume. It has been observed from simulations of confined hexadecane that the compressibility and bulk modulus of hexadecane

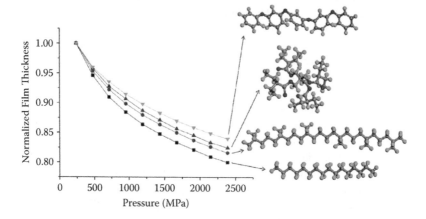

FIGURE 8.11 Normalized film thickness as a function of pressure illustrating the response of four confined fluids (from the bottom: n-hexadecane, squalane, pentaerythritol tetra, and polyphenylether) to compression.

predicted by MD is only slightly less (~2%) than that predicted using standard empirical models (Dowson-Higginson at low pressure, Tait at high pressures) (Martini and Vadakkepatt 2010). However, to this point, the comparison between thin film and bulk compressibility has not been made for a sufficiently large set of fluids or operation conditions to support a general statement about their consistency.

Another effect that can be significant for fluids confined to molecular length scales is solvation pressure. This is an additional component of pressure that acts perpendicular to the confining walls. Solvation pressure is obtained from an MD simulation by calculating the stress tensor and then taking the difference between the normal stress in the direction perpendicular to the walls and the average of the normal stresses in the other two directions. Note that stress calculated from an MD simulation is usually the virial stress, which is a measure of momentum change due to interatomic forces and mass transport. While it is not equivalent to mechanical Cauchy stress (Zhou 2003), its distribution does provide a qualitative understanding of how the stress varies in the system over time and space. Solvation pressure is observed to oscillate with channel width, as shown in Figure 8.12 for n-decane. The periodicity of the oscillations is comparable to the characteristic length scale of the fluid molecules, and the oscillation amplitude decays rapidly with increasing channel width. In the case of linear hydrocarbons such as n-decane, this characteristic length is approximately equal to the size of a methyl group. Solvation pressure is thought to arise when the ordering of fluid molecules into discrete layers is disrupted due to confinement (Israelachvili 1992). It has also been observed that, for small channel widths at which solvation pressure is significant, the dominant contribution to the solvation force is the interaction between a wall and the fluid layer immediately next to it (Tarazona and Vicente 1985). This interaction force is necessarily a function of the separation distance between the wall and fluid layer. The distance between the wall and the first fluid layer, r, can be calculated from the distance between the wall and the first peak in the atomic density profile. At large channel widths,

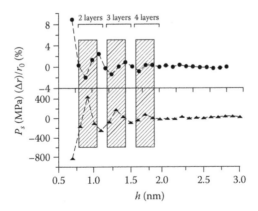

FIGURE 8.12 Average difference between the wall–fluid layer distance and the equilibrium distance (top) and the solvation pressure (bottom) as functions of channel width. Periodicity common to both wall-fluid layer distance and solvation pressure is highlighted by shaded bars.

h, this layer distance reaches a constant, equilibrium value, $r(h{\to}\infty) = r_0$. However, for fluids confined to a small number of layers, the wall–fluid layer distance varies with channel width. The difference between the wall–fluid layer distance at a given channel width and the equilibrium distance can be defined $\Delta r = r - r_0$. The percent difference from equilibrium ($\Delta r / r_0$) oscillates with channel width between negative, where the wall–layer distance is smaller than the equilibrium value, and positive, where the wall–layer distance is larger than the equilibrium value. This periodic wall–fluid layer distance is observed to be inversely correlated with the oscillatory behavior of solvation pressure as shown in Figure 8.12. This relationship can be understood as follows: If the distance between the wall and the first liquid layer is less than the equilibrium value, the fluid molecules exert a force on the walls that tends to push them apart (positive solvation pressure). If the wall–fluid layer distance is larger than the equilibrium value, the fluid molecules exert a force that tends to pull the walls together (negative solvation pressure). It is important to point out that solvation pressure oscillations can be expected to decrease with increasing chain length and branching in chain molecules, and with increasing wall surface roughness (Porcheron et al. 2002).

The last two properties that will be discussed are viscosity (a measure of resistance to shear) and interface slip (a measure of the difference between velocity of a solid wall and that of the adjacent fluid). Unlike density and solvation pressure, these properties are functions of shear rate and, as with the velocity in atomic-scale friction, shear rates accessible to MD simulation are typically many orders of magnitude larger than those available to experimentalists. Further, the temporal parallelization methods that can be applied to solid material models cannot be applied to fluids because the quasi-static assumption is no longer applicable. As a result, the predictions of viscosity and interface slip obtained from most MD simulations must be extrapolated to lower (more physically realistic) shear rates. The calculation methods for viscosity and slip are similar in the sense that they both rely on the fluid velocity profile. A representative velocity profile for n-hexadecane in Couette flow is shown in Figure 8.13. Velocity profiles are obtained by grouping the atoms into bins, as is done

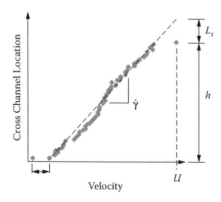

FIGURE 8.13 Velocity profile of a fluid in Couette flow illustrating the variables used in calculation of viscosity and interface slip from MD simulation.

to calculate density profiles, and then averaging the velocity of the atoms in each bin over time. A linear fit to the average velocity of the bins in the center of the channel is the shear rate, $\dot{\gamma}$. This shear rate is sometimes called the *actual shear rate* to differentiate it from the *effective shear rate*, which is calculated as the relative velocity of the two walls divided by the distance between them. The viscosity is calculated as the shear stress (typically obtained using the virial stress) divided by the shear rate (actual and effective viscosities are calculated using the actual and effective shear rates, respectively). Interface slip can be quantified in two ways: slip velocity, u_s, or slip length, L_s. As illustrated in Figure 8.13, slip velocity is the difference between the wall velocity and the velocity of the first fluid layer, and slip length is the distance from the wall at which the extrapolated shear rate will reach the wall velocity. Note that slip length and slip velocity are necessarily related via the shear rate. Shear can also be imposed on a fluid by application of a pressure gradient (Poiseuille flow). In that case, the calculation of slip velocity is the same—the difference between the wall speed (zero) and the velocity of the first fluid layer. However, calculation of the shear rate differs in that the velocity profile is parabolic instead of linear.

MD is often used to study the effect of shear rate on viscosity and interface slip. However, it is important to note up front that most of those studies observe both viscosity and slip predictions to be significantly affected by the atomic interaction model (particularly that between the wall and fluid atoms) (Thompson and Robbins 1990). An illustration of the effect of wall–fluid interaction strength on slip length is shown for several fluids in Figure 8.14. Another critical point is that shear rate can be varied by simulating constant film thickness and variable wall speed (or pressure gradient), or variable film thickness and constant wall speed (or pressure gradient). These two approaches may lead to very different results because both viscosity and interface slip can be affected by confinement (i.e., vary with film thickness) (Jabbarzadeh et al. 1998, Martini et al. 2006). In MD-based studies of viscosity, the most commonly reported behavior is shear thinning. Shear thinning, the decrease of viscosity with increasing shear rate, is not specific to thin films. However, it is particularly important in thin film rheology studies because thin film lubricants are subject to large shear rates (often above the limit at which shear thinning occurs), even

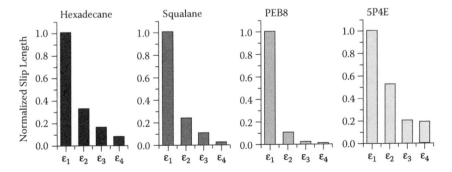

FIGURE 8.14 Effect of wall–fluid interaction strength (increasing from ε_1 to ε_4) on slip length for four different fluids. Ordinate axes are normalized by slip length at the smallest wall–fluid interaction strength for each fluid.

under moderate operating speeds. Although the mechanisms underlying shear thinning on different length scales may not be the same, viscosity–shear rate data have been found to follow consistent, length scale–independent behavior using the time temperature superposition principle (Bair et al. 2002). Reported shear rate effects on slip length are much less consistent than those for viscosity. Not only are reported trends very different (especially at high shear rates), but authors interpret and explain their results using very different arguments. Results are difficult to compare from one study to the next because even small differences between models can lead to significant variation in slip length. As mentioned previously, the most critical model parameter appears to be the wall–fluid interaction strength. However, it has also been shown that surface roughness and flexibility can influence model predictions (Priezjev and Troian 2006). Wall flexibility in this case refers to the ability of the wall atoms to move and so enable momentum transfer. It has been shown that models with rigid wall atoms (fixed in space) will predict that slip length increases rapidly at high shear rates, while models that allow momentum transfer will predict that slip length asymptotes to a constant value at high shear rates (Martini et al. 2008).

8.4 SUMMARY

In summary, MD simulation is a research tool that can be used to investigate tribological phenomena on length scales where discrete atomic events are significant. All aspects of nanotribology are at some fundamental level influenced by atomic interactions. As this chapter has illustrated, researchers have already employed molecular dynamics simulation to gain insight into the fundamental mechanisms of atomic-scale friction and thin film lubrication. The ability of this powerful modeling tool to contribute to the field of tribology can be expected to expand dramatically as the simulations themselves improve (e.g., better empirical interaction models and faster computational resources) and the need for atomic-scale research increases (e.g., smaller components and tighter tolerances). These trends suggest that future molecular dynamics simulation research can contribute significantly not only to our understanding of atomic-scale phenomena, but the ability to predict the behavior of tribological interfaces.

ACKNOWLEDGMENTS

This chapter could not have been written without the significant contributions of Yalin Dong and Ajay Vadakkepatt.

REFERENCES

Bair, S., McCabe, C., and Cummings, P.T. 2002. Comparison of nonequilibrium molecular dynamics with experimental measurements in the nonlinear shear-thinning regime. *Physical Review Letters* 88: 058302.

Gao, G., Cannara, R.J., Carpick, R.W., and Harrison, J.A. 2007. Atomic-scale friction on diamond: A comparison of different sliding directions on (001) and (111) surfaces using MD and AFM. *Langmuir* 23: 5394–5405.

Gao, G., Mikulski, P.T., and Harrison, J.A. 2002. Molecular-scale tribology of amorphous carbon coatings: Effects of film thickness, adhesion, and long-range interactions. *J. Am. Chem. Soc.* 124: 7202–7209.

Gupta, S.A., Cochran, H.D., and Cummings, P.T. 1997. Shear behavior of squalane and tetracosane under extreme confinement. II. Confined film structure. *Journal of Chemical Physics* 107: 10327–10334.

Hu, Y.-Z., and Granick, S. 1998. Microscopic study of thin film lubrication and its contributions to macroscopic tribology. *Tribology Letters* 5: 81–88.

Israelachvili, J.N. 1992. *Intermolecular and surface forces.* San Diego: Academic Press.

Jabbarzadeh, A., Atkinson, J.D., and Tanner, R.I. 1998. Nanorheology of molecularly thin films of n-hexadecane in Couette shear flow by molecular dynamics simulation. *J. Non-Newtonian Fluid Mech.* 77: 53–78.

Kim, W.K., and Falk, M.L. 2010. Accelerated molecular dynamics simulation of low-velocity frictional sliding. *Modelling and Simulation in Materials Science and Engineering* 18: 034003.

Landman, U., Luedtke, W.D., and Ringer, E.M. 1992. *Wear* 153: 3.

Li, Q., Dong, Y., Perez, D., Martini, A. and Carpick, R. W. 2011. Speed dependence of atomic stick-slip friction in optimally matched experiments and molecular dynamics simulations: the role of dynamics vs. energetics. *Phys. Rev. Letters* 106, 126101.

Luan, B., and Robbins, M.O. 2005. The breakdown of continuum models for mechanical contacts. *Nature Letters* 435: 929–932.

Martini, A., Dong, Y., Perez, D., and Voter, A.F. 2009. *Tribol. Lett.* 36: 63.

Martini, A., Hsu, H.-Y., Patankar, N.A., and Lichter, S. 2008. *Phys Rev Lett* 100: 206001.

Martini, A., Liu, Y.C., Snurr, R.Q., and Wang, Q. 2006. Molecular dynamics characterization of thin film viscosity for EHL simulation. *Tribology Letters* 21: 217–225.

Martini, A., and Vaddakepatt, A. 2010. Compressibility of thin film lubricants characterized using atomistic simulation. *Tribol Lett* 38: 33.

Mishin, Y., Suzuki, A., Uberuaga, B.P., and Voter, A.F. 2007. Stick-slip behavior of grain boundaries studied by accelerated molecular dynamics. *Phys. Rev. B* 75: 224101.

Mo Y., Turner, K.T., and Szlufarska, I. 2009. Friction laws at the nanoscale. *Nature Letters* 457: 1116.

Müser, M.A., and Robbins, M.O. 2000. Conditions for static friction between flat crystalline surfaces. *Phys Rev B* 61: 2335.

Nieminen, J.A., Sutton, A.P., and Pethica, J.B. 1992. Static junction growth during frictional sliding of metals. *Acta. Metall.* 40: 2503–2509.

Perez, D., Dong, Y., Martini, A., and Voter, A.F. 2010. Rate theory description of atomic stick-slip friction. *Phys. Rev. B* 81: 245415.

Porcheron, F., Rousseau, B., and Fuchs, A.H. 2002. Structure of ultra-thin confined alkane films from Monte Carlo simulations. *Molecular Physics* 100: 2109.

Priezjev, N.V., and Troian, S.M. 2006. Influence of periodic wall roughness on the slip behaviour at liquid/solid interfaces: Molecular-scale simulations versus continuum predictions. *J. Fluid Mech.* 554: 25.

Robbins, M.O., and Smith, E.D. 1996. Connecting molecular-scale and macroscopic tribology. *Langmuir* 12: 4543.

Sørensen, M.R., Jacobsen, K.W., and Stoltze, P. 1996. Simulations of atomic-scale sliding friction. *Phys. Rev. B* 53: 2101.

Snook, I., and van Megen, W. 1979. Structure of dense liquids at solid interfaces. *Journal of Chemical Physics* 70: 3099.

Szlufarska, I., Chandross, M., and Carpick, R.W. 2008. Recent advances in single-asperity nanotribology. *J. Phys. D: Appl. Phys.* 41: 123001.

Tarazona, P., and Vicente, L. 1985. A model for density oscillations in liquids between solid walls. *Molecular Physics* 56: 557.

Thompson, P.A., and Robbins, M.O. 1990. Shear flow near solids: Epitaxial order and flow boundary conditions. *Phys Rev A* 41: 6830.

Voter, A.F. 1998. Parallel replica method for dynamics of infrequent events. *Phys. Rev. B* 57: 13985.

Wenning, L., and Müser, M.H. 2001. Friction laws for elastic nanoscale contacts. *Europhys. Lett.* 54: 693.

Zhou, M. 2003. A new look at the atomic level virial stress: On continuum-molecular system equivalence. *Proc. R. Soc. Lond. A* 459: 2347.

Index